MATH IS FUN 5A

For Grades 5-8

Solve Engaging Puzzles and Learn Mathematics

Panna Lal Patodia

Copyright ©2018 Panna Lal Patodia

Dedicated to my wife and kids for their love and support.

MATH IS FUN 5A

Preface

I am a software engineer and not a teacher. Recently, I had an opportunity to teach mathematics to a class of seventh grade students. The number of students were around 30. Out of them, few were brilliant, others were mediocre, and rest were not interested in mathematics. The key challenge was to teach mathematics to the 10 students who were having a kind of fear for mathematics.

When all my efforts failed, and I was on the verge of giving it up, I just put a puzzle to all the students. It was a simple puzzle. Below, I give the puzzle. Please see the Figure 1 on the next page and you will see the puzzle. The challenge was to insert the numbers 2, 3, 6, 8 and 9 in appropriate place so that the six equations (three rows and three columns) are satisfied.

Once the students were explained what they were supposed to do, they started inserting numbers using trial and error. For example, if we insert number 2 in the first column of the first row, the sum of the first column will be 7 but it is supposed to be 14. Thus, the number required was bigger. If we put 8 instead of 2, we get 8 + 1 + 4 = 13 but we still did not get 14. So, we put 9 and check 9 + 1 + 4 = 14 and now the sum is equal to 14.

Next, we take the third row. In third row, if we put 2 in the third column of the third row, we get 4 x 5 x 2 = 40 and this is equal to the result shown. Thus, the value 2 is the right one.

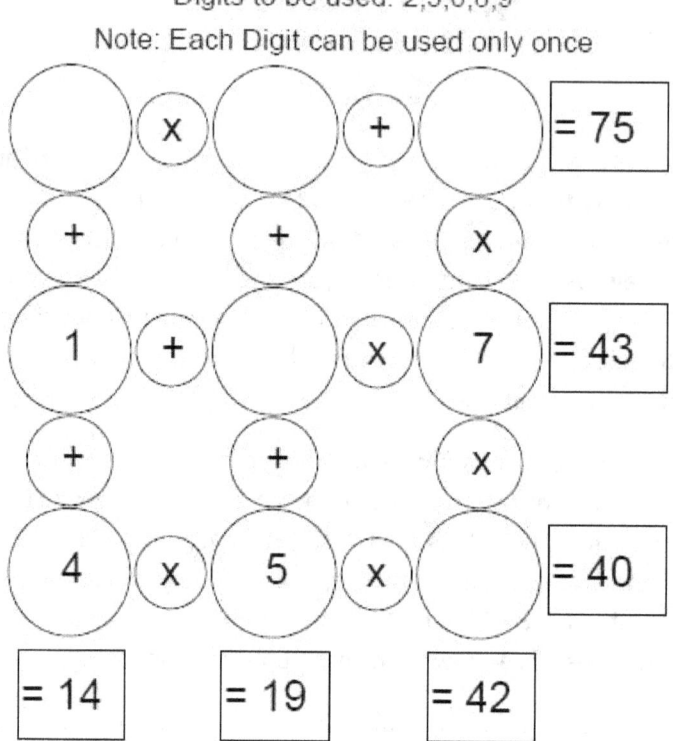

Digits to be used: 2,3,6,8,9
Note: Each Digit can be used only once

Figure 1

In this way, using trial and error, we can fill all the five numbers.

Initially, some of the students faced problems. However, soon, they overcome these problems and started enjoying.

I made groups wherein each group had 5 students. The challenge was to finish the puzzle correctly and in minimum time. Each member of the group who finished first was given a reward. Soon, all the students found these puzzles very interesting and they were fully absorbed in solving these puzzles. The reason of their excitement and involvement was

(1) They were no more solving mathematical problems, rather they were playing.
(2) There was a kind of competition among students to finish the puzzle first.

I have designed number of such puzzles and I was giving every day new puzzle to students. We reserved last 10 minutes of the mathematics period for this activity.

While solving these puzzles, the students were not learning addition, subtraction, multiplication and division, they were also learning equation solving skills indirectly.

Almost all the students have shown remarkable improvements and keen interest in mathematics once they started solving these puzzles. So, I thought to share these puzzles by publishing this book so that all the students in the world can be benefited.

This book contains 100 puzzles with solution. Each puzzle has unique solution. Based on the response from our readers, I shall come out with lot more such books so that students can solve these types of puzzles and overcome the fear of mathematics.

Contents

MATH IS FUN 5A

Introduction

Let us start with a sample puzzle given below:

Digits to be used: 2,3,6,8,9

Note: Each Digit can be used only once

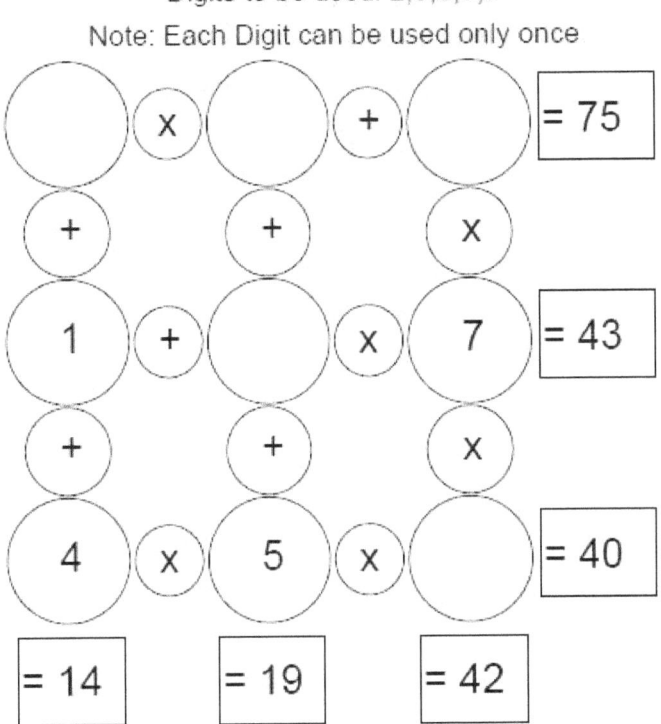

Figure 1

Let us understand the problem. Each problem has three rows and three columns. Four numbers are already inserted and we have to insert digits 2, 3, 6, 8 and 9 in such a manner so that all six equations (three rows and three columns) are satisfied.

Let us solve the above problem. There are three rows and three columns. Values are given for three rows and three columns. Let us start with the first column. Please note that we can solve the problem using different methods. Below, we shall explore different methods.

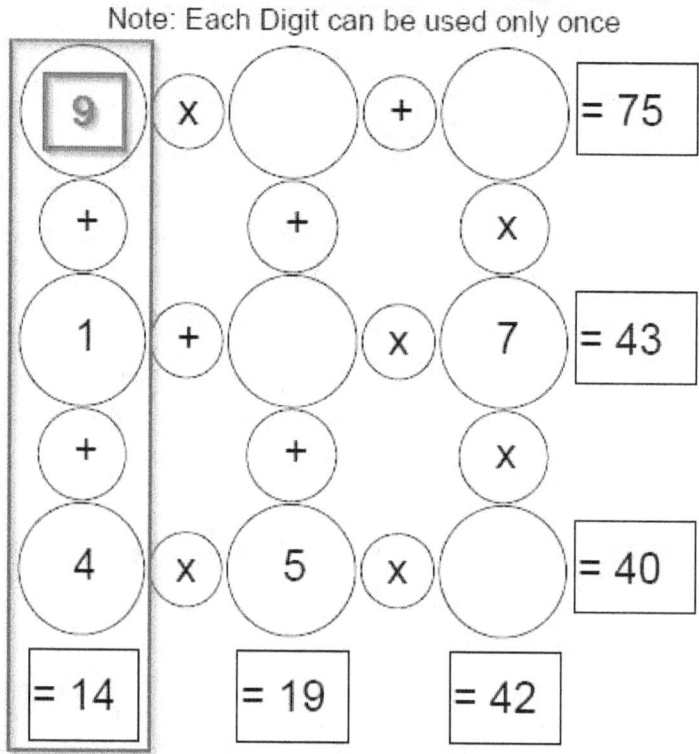

Digits to be used: 2,3,6,8,9
Note: Each Digit can be used only once

Figure 2

[2]

First Number (See Figure 2):

Let us take the first column and find the missing number. We shall solve the problem using different methods:

(a) **Trial and Error**: Suppose, we insert number 2 in the first column of the first row, the sum of the first column will be 7 but it is supposed to be 14. Thus, the number required was bigger. If we put 8 instead of 2, we get 8 + 1 + 4 = 13 but we still did not get 14. So, we put 9 and check 9 + 1 + 4 = 14 and now the sum is equal to 14.

(b) **Arithmetical Method:** When we add a number to 1 and 4, we get 14. What is the number? Answer is simple: 14 − (4+1) = 14 − 5 = 9.

(c) **Algebraic Method:** We can also find the missing number by solving equation:

$$x + 1 + 4 = 14 \text{ or } x + 5 = 14 \text{ or } x = 9$$

(d) **Word Problem:** We can frame the question:
William has 1+4 = 5 mangoes. If he got some mangoes from his mother and now, he has 14 mangoes. How many mangoes was given by his mother. It is simple to solve. Just, subtract 5 from 14 and the answer is 9. So, we should insert 9 in the first column of the first row.

Second Number (See Figure 3):

Let us take now third row and find the missing number.

(a) Trial and Error: Now, we are left with the numbers 2, 3, 6 and 8. Let us try 2. If we put 2 in the third column of the third row, we get 4 x 5 x 2 = 40 and this is equal to the result shown. Thus, the value 2 is the right one.

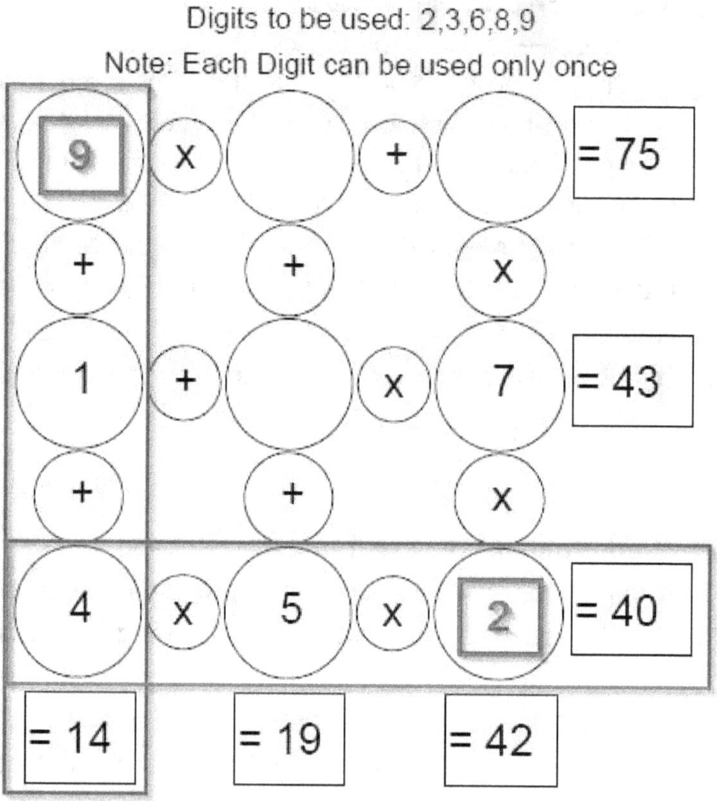
Digits to be used: 2,3,6,8,9
Note: Each Digit can be used only once
9
x
+
= 75
+
+
x
1
+
x
7
= 43
+
+
x
4
x
5
x
2
= 40
= 14
= 19
= 42

Figure 3

(b) Arithmetical Method: Find a number when it is multiplied with 4 x 5 gives the result 40. To get the answer, we divide 40 by (4 x 5) or 40 by 20. The result is 2.

(c) **Algebraic Method:** We can also compute the number by solving the equation:

$$4 \times 5 \times x = 40 \ or \ 20x = 40. \ So, x = \frac{40}{20} = 2$$

(d) **Word Problem:** Let us frame the above problem in a different way. Sam has 40 apples. If he wants to distribute these apples among 20 of his friends, how many apples each friend will get. On dividing 40 by 20, we get 2. So, we can put 2 in the third column of the third row.

Third Number (See Figure 4): Now, let us take second row and find the missing digit.

(a) **Trial and Error:** Now, we are left with three digits 3, 6 and 8. Let us see what happens if we put 3. The result is 1 + 3 x 7 = 22. However, we want the result to be 43. Let us try next number 6. When, we insert 6, we get 1 + 6 x 7 = 43. This perfectly matches with the given value. So, the answer is 6.

(b) **Arithmetical Method:** The question is to find a number when multiplied by 7 and added with 1, we get 43. To get the answer we divide (43-1) by 7 or 42 by 7. The answer is 6.

(c) **Algebraic Method:** We can also write this as equation and solve:

$$1 + x \times 7 = 43 \ or \ 7x = 42 \ or \ x = \frac{42}{7} = 6$$

Digits to be used: 2,3,6,8,9

Note: Each Digit can be used only once

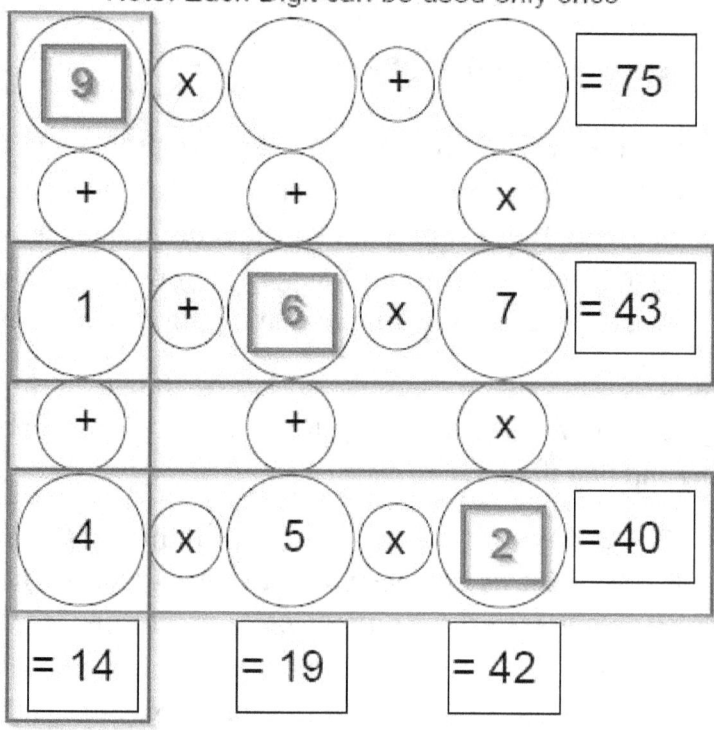

Figure 4

(d) **Word Problem:** Alternatively, we can frame the question: Sophia has 43 oranges, she has eaten 1 orange and rest of the oranges were equally distributed among her 7 friends. On subtracting 1 from 43, we get 42. If we distribute 42 oranges among 7 friends, each friend will get

6 oranges. So, let us put 6 in the second column of the second row.

<u>Fourth Number</u> (See Figure 5):

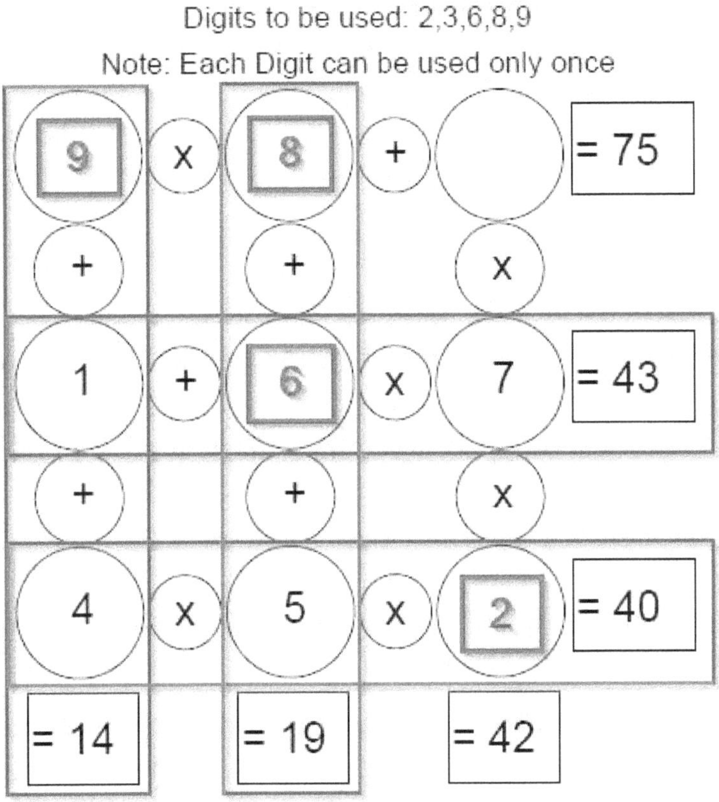
Digits to be used: 2,3,6,8,9
Note: Each Digit can be used only once
9
x
8
+
= 75
+
+
x
1
+
6
x
7
= 43
+
+
x
4
x
5
x
2
= 40
= 14
= 19
= 42

Figure 5

Now, let us take the second column and find the missing digit.

(a) Trial and Error: Now, we are left with two numbers 3 and 8. If we insert number 3, we get the result 3 + 6 + 5 =

14 whereas the result shown is 19. So, let us try another number 8. On putting 8, we get 8 + 6 + 5 = 19. This matches with the result given. So, the answer is 8.

(b) Arithmetical Method: When we add a number to 6 and 5, we get 19. Find the number. We can compute the number by subtracting 6+5 from 19. The answer is 19 − (6 + 5) = 19 − 11 = 8.

(c) Algebraic Method: we can also compute the number by solving the equation:

$$x + 6 + 5 = 19 \; or \; x + 11 = 19 \; or \; x = 8$$

So, we can put 8 in the second column of the first row.

(d) Word Problem: Emily has 11 cherries. Her father gave some more cherries to her. Now, she has 19 cherries. How many cherries are given by her father? The answer is simple 19-11 = 8 cherries.

Fifth Number (See Figure 6): Now, only one circle is blank that is in the third column of first row.

(a) Trial and Error: Now, we are left with only one number and that is 3. Let us put 3 and check the row and column. For first row, 9 x 8 + 3 = 75 that matches perfectly. If we check the third column, we get 3 x 7 x 2 = 42. This also matches. Thus, the answer is 3.

(b) Arithmetical Method: Compute a number that becomes 42 when multiplied by 7 x 2. To compute the number, we divide 42 by (7 x 2) or 42 by 14 and get the answer 3.

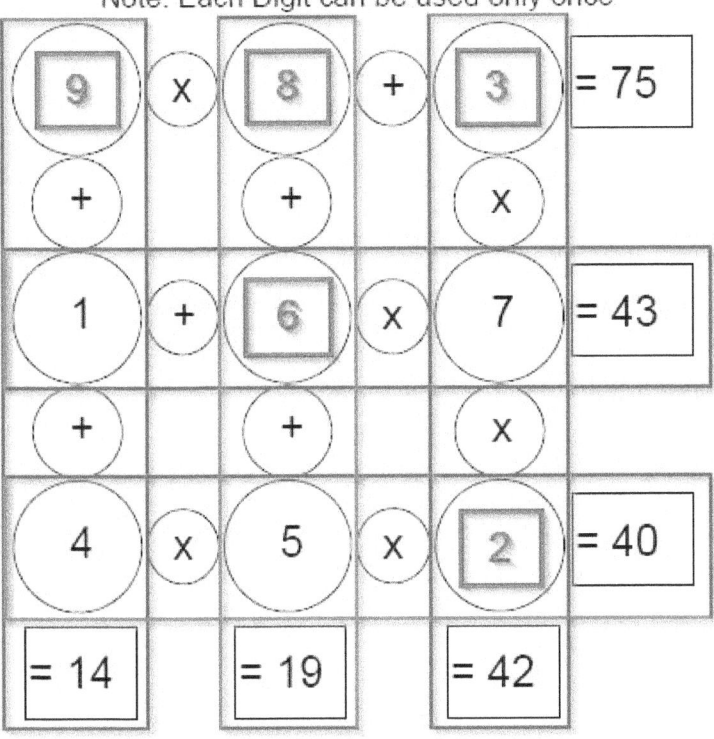

Digits to be used: 2,3,6,8,9
Note: Each Digit can be used only once

Figure 6

(c) Algebraic Method: We can also compute the number by solving the following equation:

$$x \times 7 \times 2 = 42 \; or \; 14x = 42 \; or \; x = \frac{42}{14} = 3$$

(d) Word Problem: Lilly has 75 cookies. She has given 9 cookies to each of her 8 friends. How many cookies are left with her? The answer is simple 75 – 9 x 8 = 75 -72 = 3 cookies.

We have solved the above puzzle by using four different methods. Students can use any method of their choice. Even while solving a puzzle, they can use different method to different missing numbers.

This book has 100 such interesting problems. I invite students to solve these problems. I am sure you will enjoy solving these problems. This will help you to get interest in mathematics and improve your mathematical skills.

Puzzle Number 1

Digits to be used: 1,4,7,8,9

Note: Each Digit can be used only once

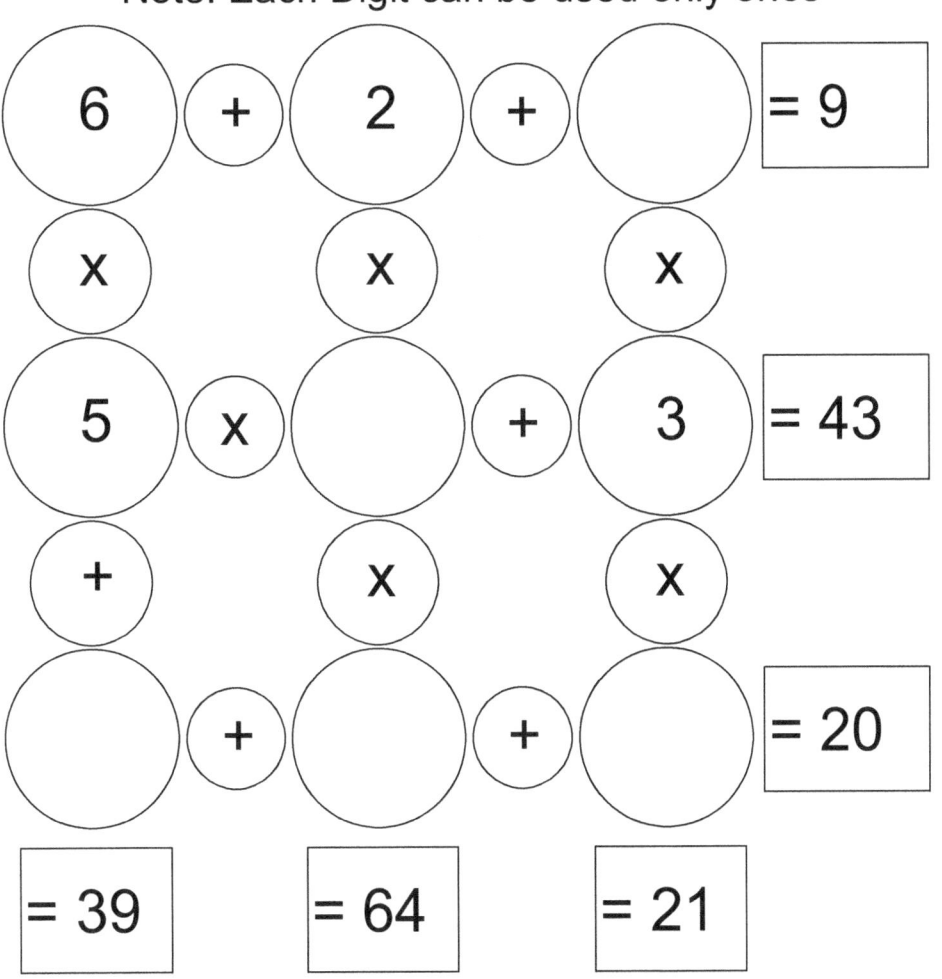

Solution at Page No 111

MATH IS FUN 5A

Puzzle Number 2

Digits to be used: 4,6,7,8,9

Note: Each Digit can be used only once

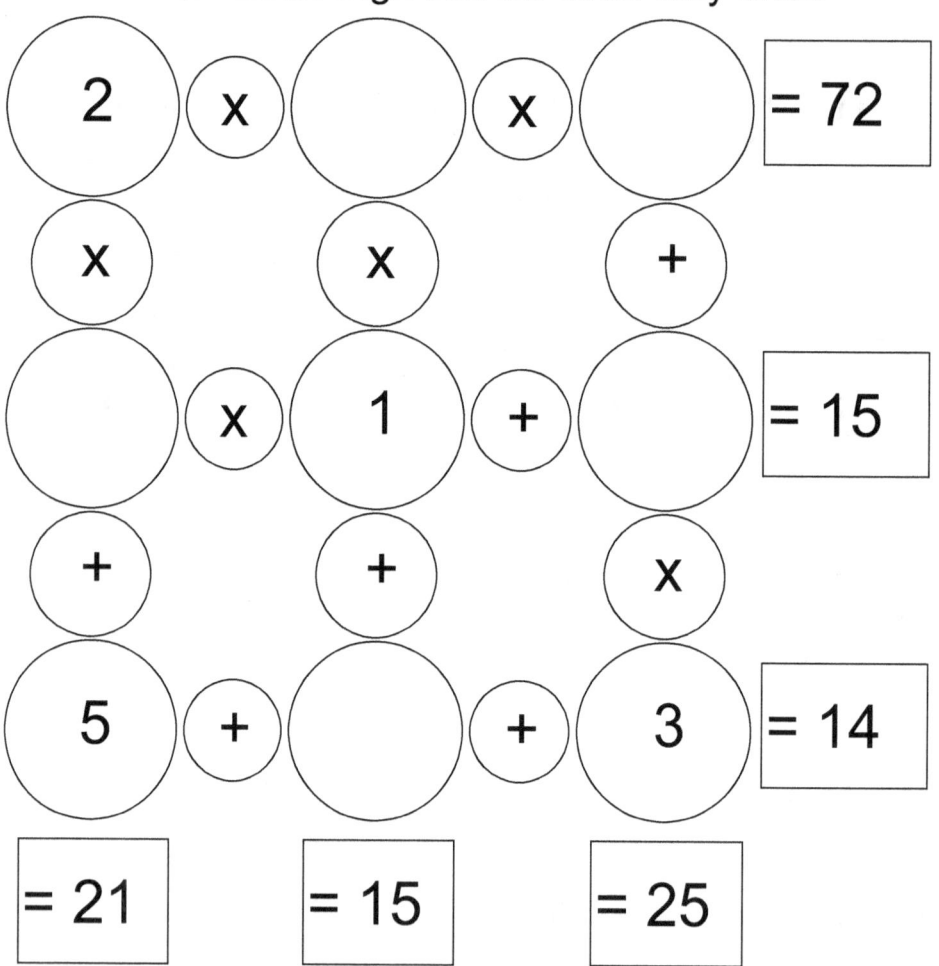

Solution at Page No 111

[12]

MATH IS FUN 5A

Puzzle Number 3

Digits to be used: 1,5,6,7,9

Note: Each Digit can be used only once

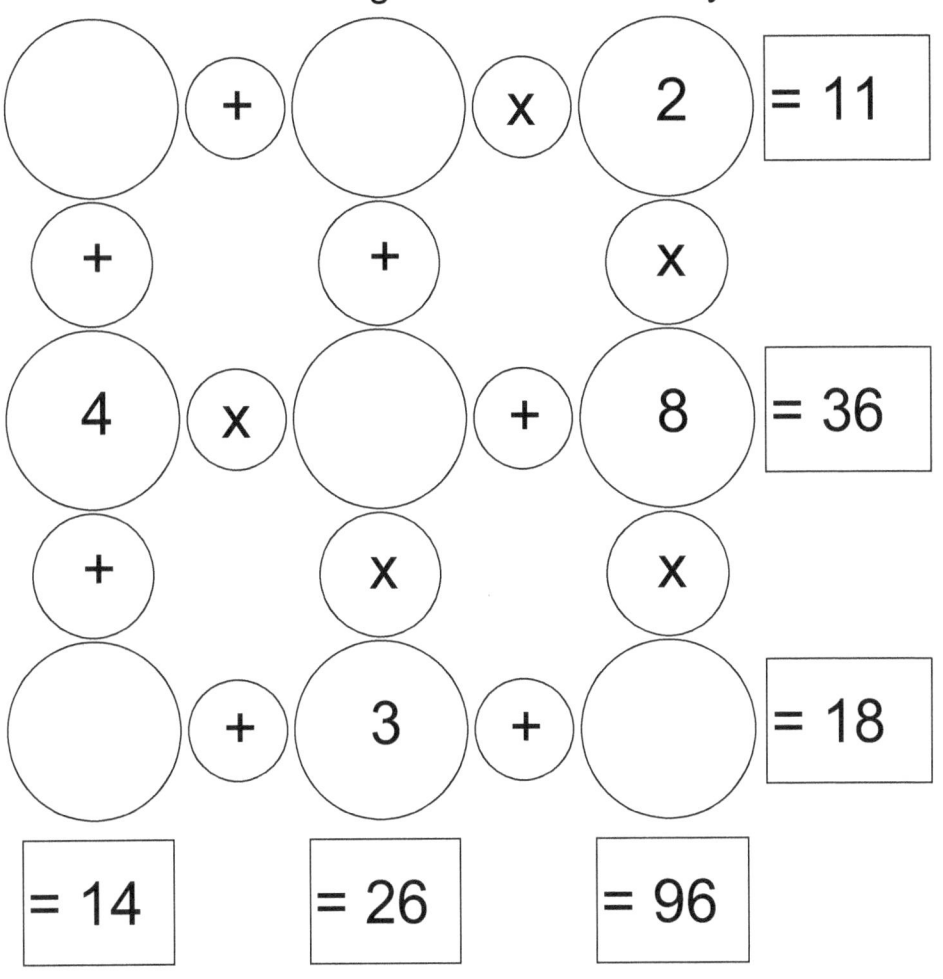

Solution at Page No 111

Puzzle Number 4

Digits to be used: 2,3,4,5,7

Note: Each Digit can be used only once

$$6 \times \underline{} + \underline{} = 46$$

$+ \qquad \times \qquad \times$

$$\underline{} \times 8 \times 1 = 40$$

$\times \qquad + \qquad +$

$$\underline{} \times 9 \times \underline{} = 54$$

$$= 16 \qquad = 65 \qquad = 7$$

Solution at Page No 111

Puzzle Number 5

Digits to be used: 2,5,6,7,8

Note: Each Digit can be used only once

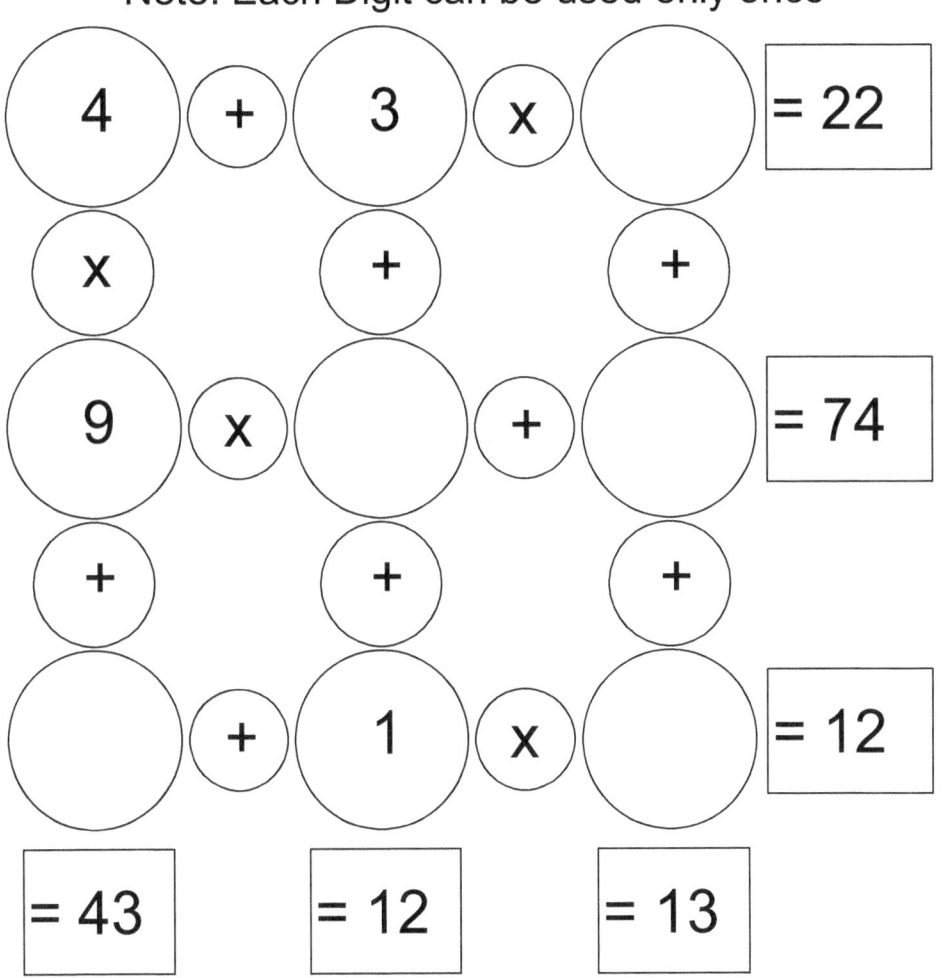

Solution at Page No 112

MATH IS FUN 5A

Puzzle Number 6

Digits to be used: 2,3,5,7,8

Note: Each Digit can be used only once

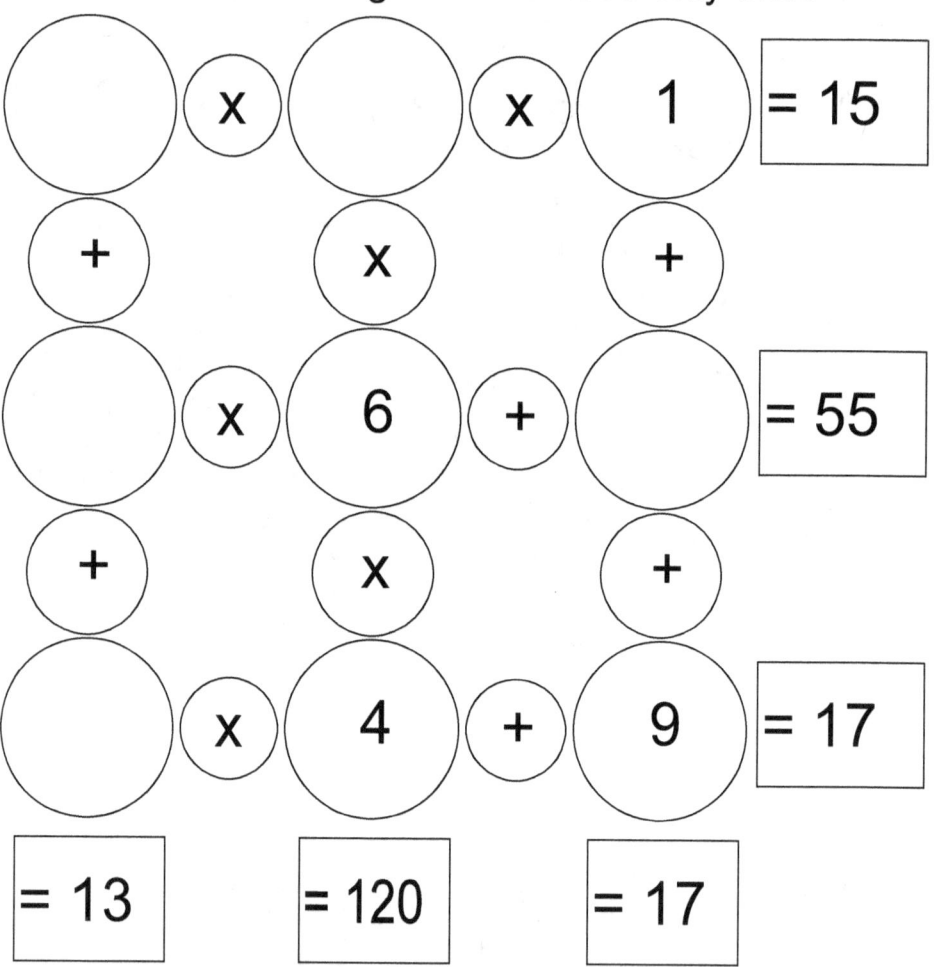

Solution at Page No 112

Puzzle Number 7

Digits to be used: 2,4,5,7,9

Note: Each Digit can be used only once

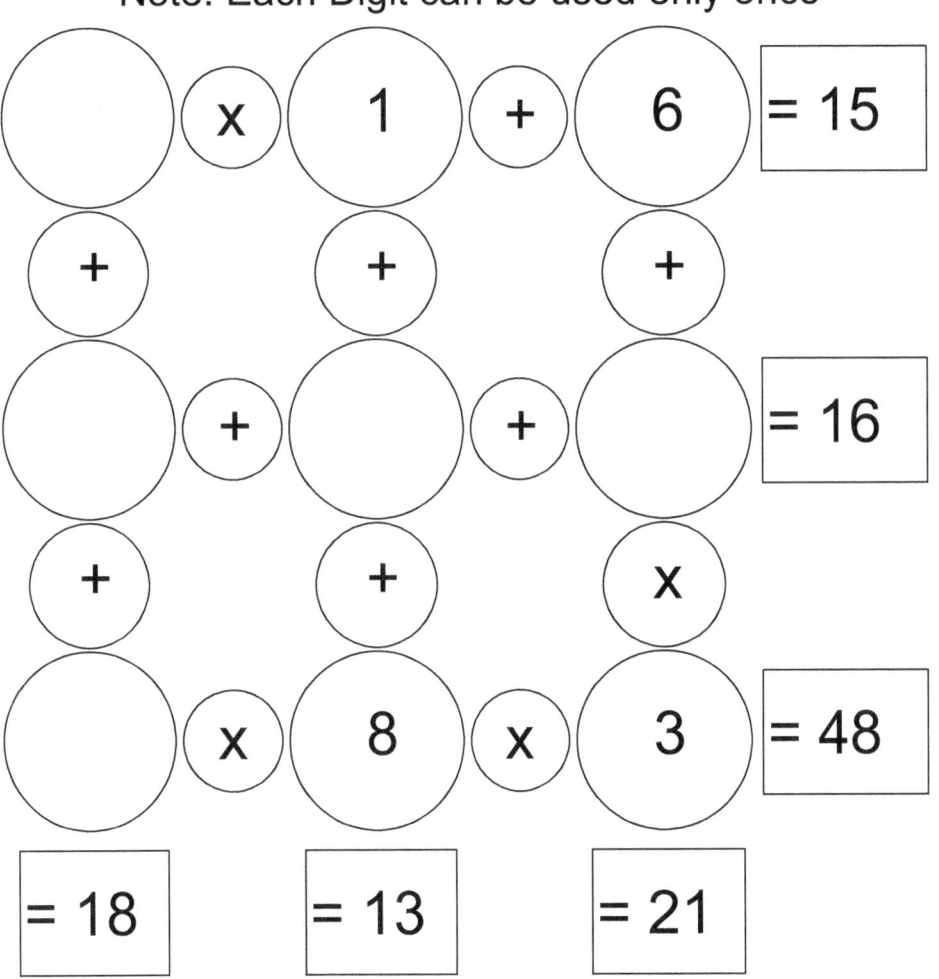

Solution at Page No 112

[17]

Puzzle Number 8

Digits to be used: 3,5,6,7,8

Note: Each Digit can be used only once

1	x	( )	x ( ) = 24

$$1 \times (\ ) \times (\ ) = 24$$

x   x   x

$$(\ ) + 4 + 9 = 19$$

+   +   +

$$(\ ) \times (\ ) \times 2 = 70$$

= 13   = 17   = 74

Solution at Page No 112

[18]

Puzzle Number 9

Digits to be used: 1,2,5,8,9

Note: Each Digit can be used only once

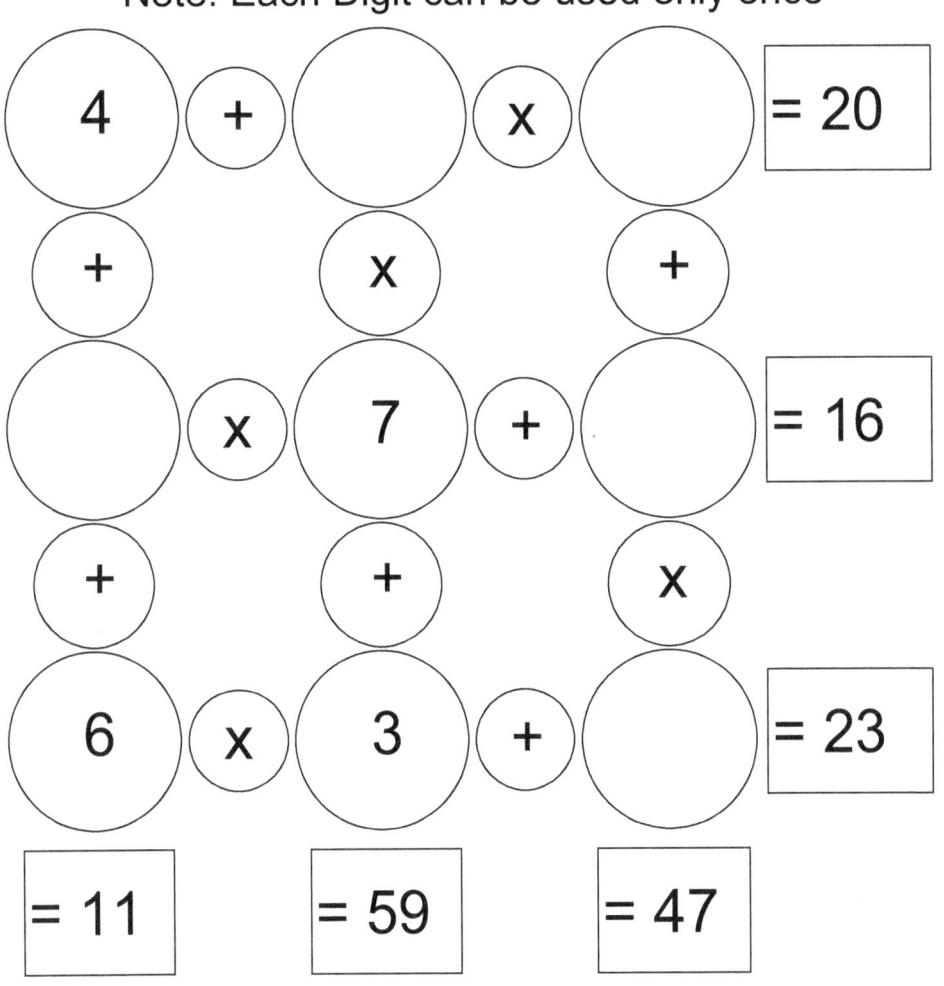

Solution at Page No 113

Puzzle Number 10

Digits to be used: 1,4,7,8,9

Note: Each Digit can be used only once

$$2 + 6 \times \bigcirc = 56$$

$$+ \quad \times \quad +$$

$$\bigcirc + 3 + \bigcirc = 15$$

$$+ \quad \times \quad +$$

$$\bigcirc \times \bigcirc \times 5 = 35$$

$$= 17 \qquad = 18 \qquad = 18$$

Solution at Page No 113

[20]

MATH IS FUN 5A

Puzzle Number 11

Digits to be used: 2,4,5,6,8

Note: Each Digit can be used only once

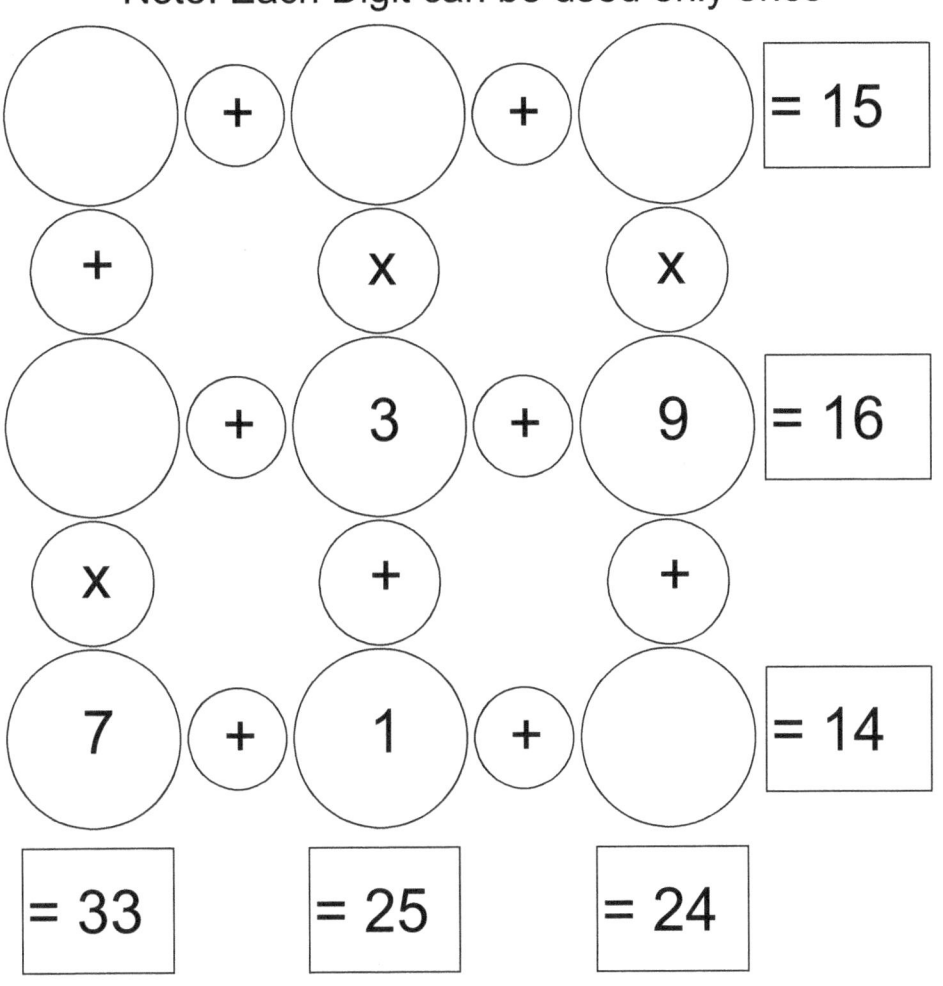

Solution at Page No 113

MATH IS FUN 5A

Puzzle Number 12

Digits to be used: 1,3,6,8,9

Note: Each Digit can be used only once

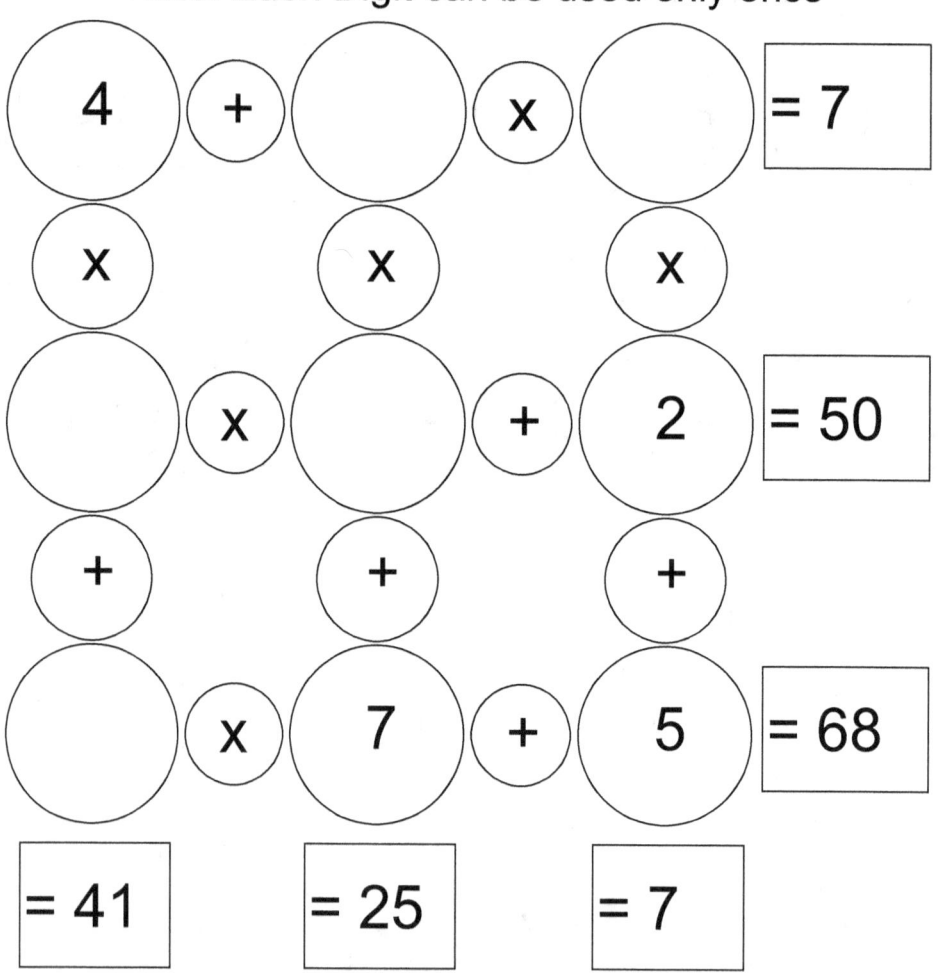

Solution at Page No 113

Puzzle Number 13

Digits to be used: 3,4,5,7,8

Note: Each Digit can be used only once

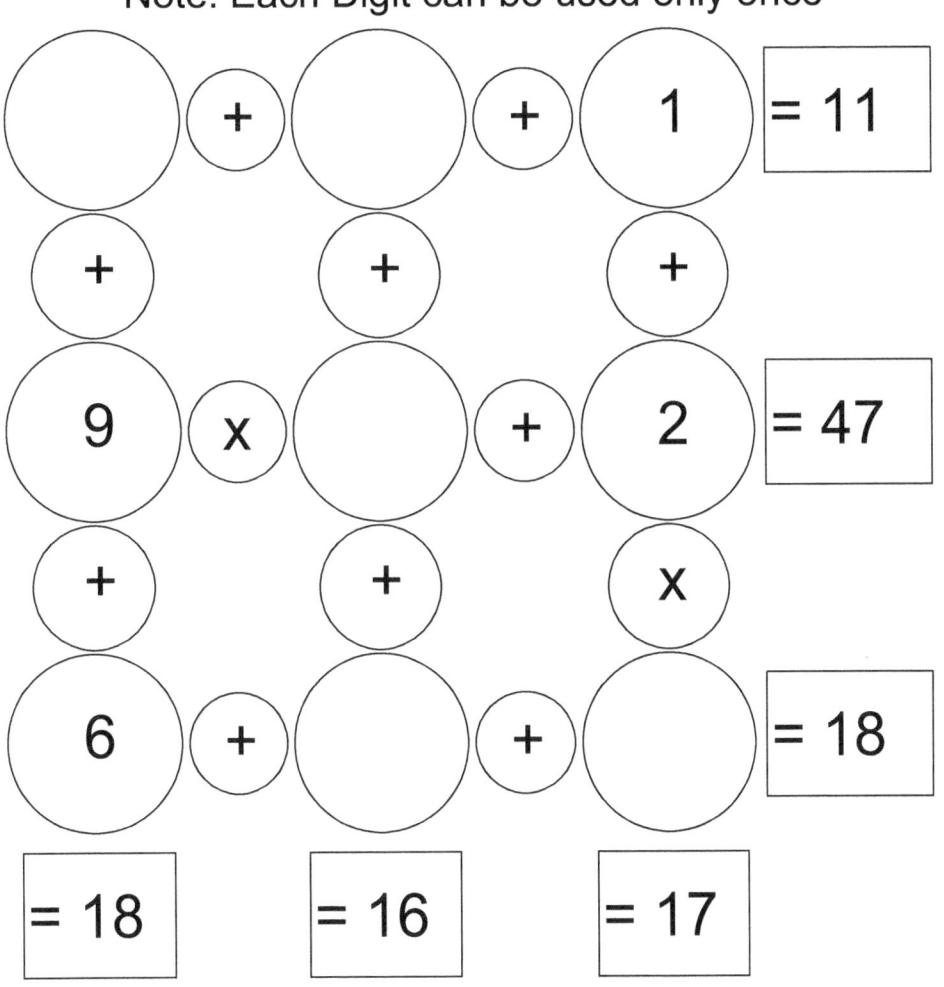

Solution at Page No 114

[23]

Puzzle Number 14

Digits to be used: 1,2,3,5,8

Note: Each Digit can be used only once

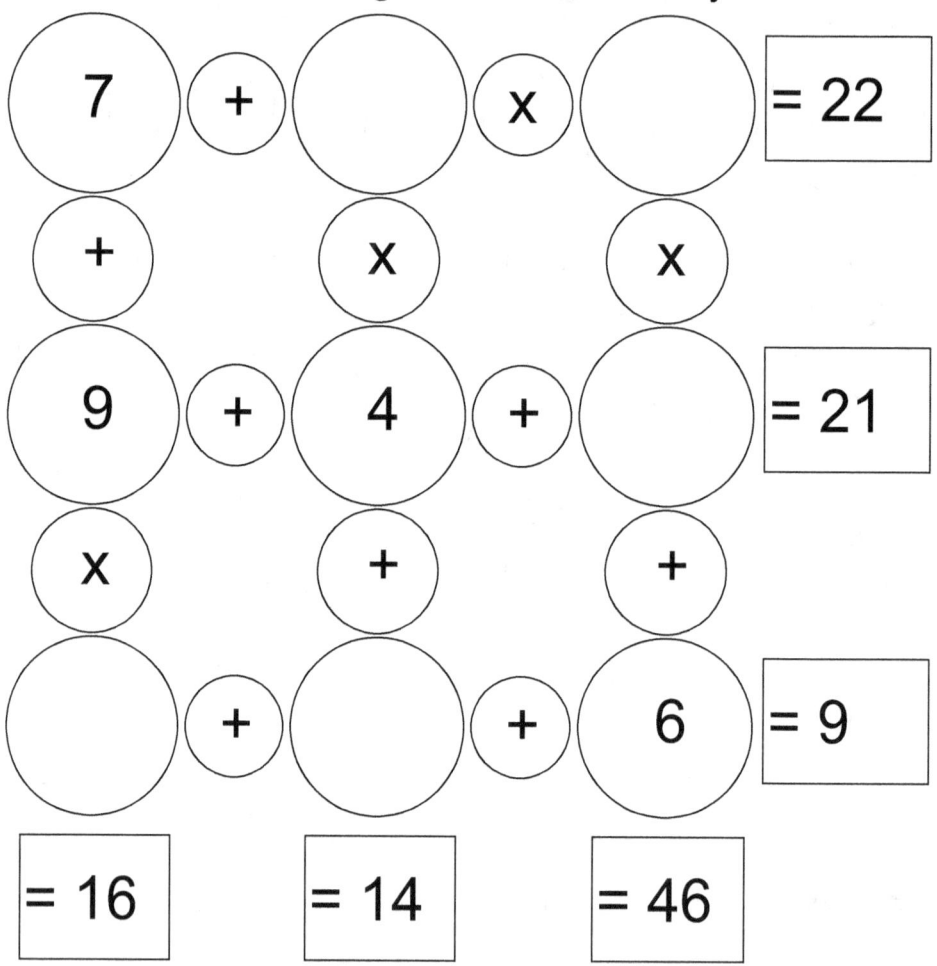

7 + ( ) x ( ) = 22

+   x   x

9 + 4 + ( ) = 21

x   +   +

( ) + ( ) + 6 = 9

= 16   = 14   = 46

Solution at Page No 114

[24]

Puzzle Number 15

Digits to be used: 2,4,5,7,8

Note: Each Digit can be used only once

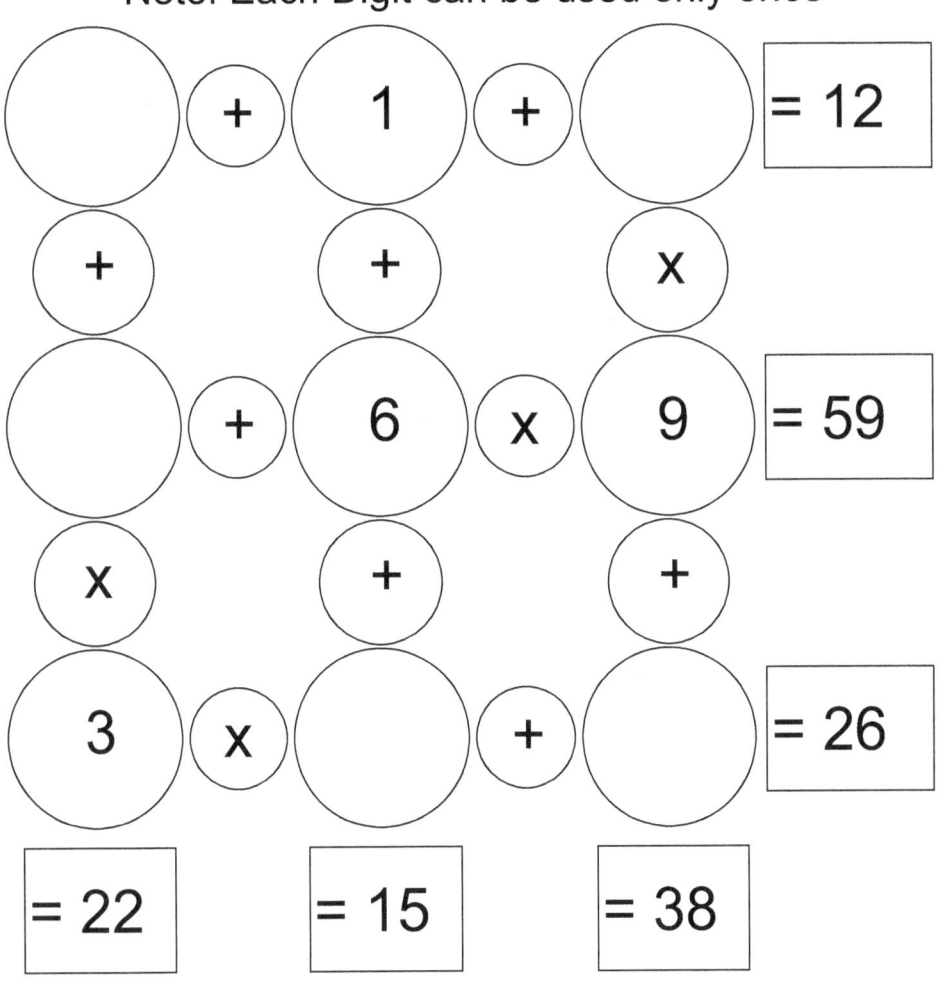

Solution at Page No 114

Puzzle Number 16

Digits to be used: 1,2,3,5,6

Note: Each Digit can be used only once

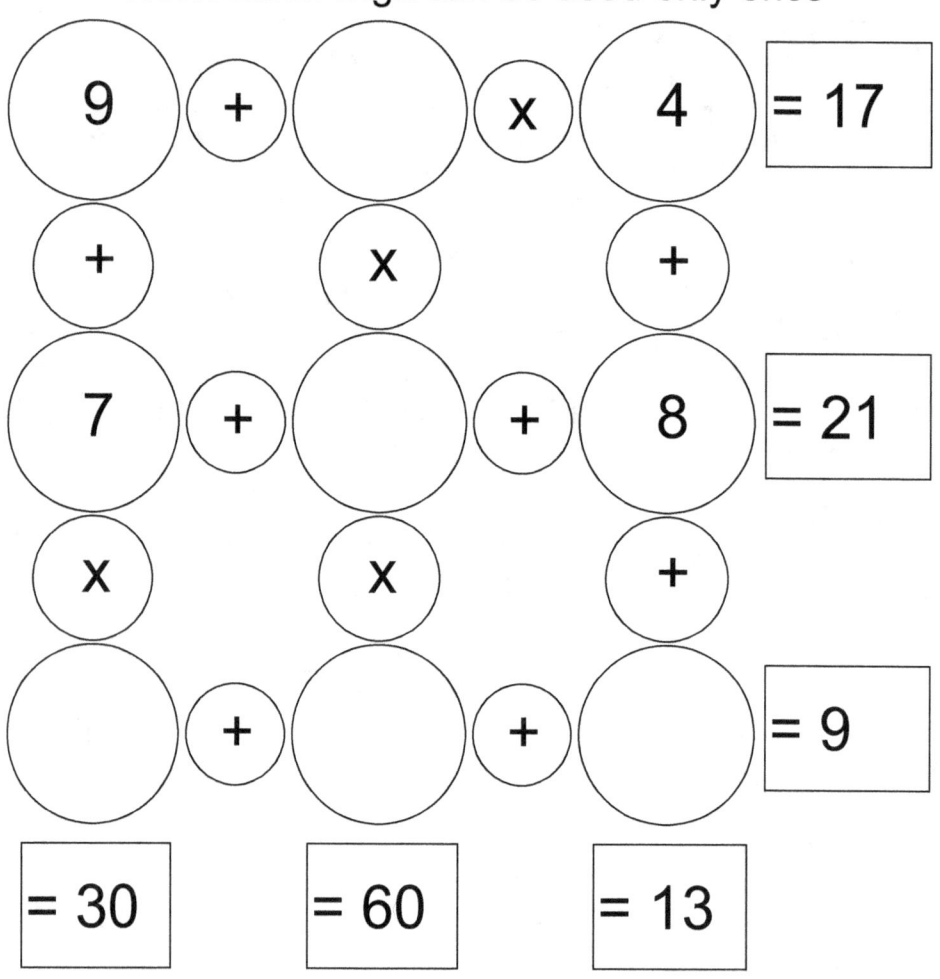

Solution at Page No 114

[26]

Puzzle Number 17

Digits to be used: 1,4,5,7,9

Note: Each Digit can be used only once

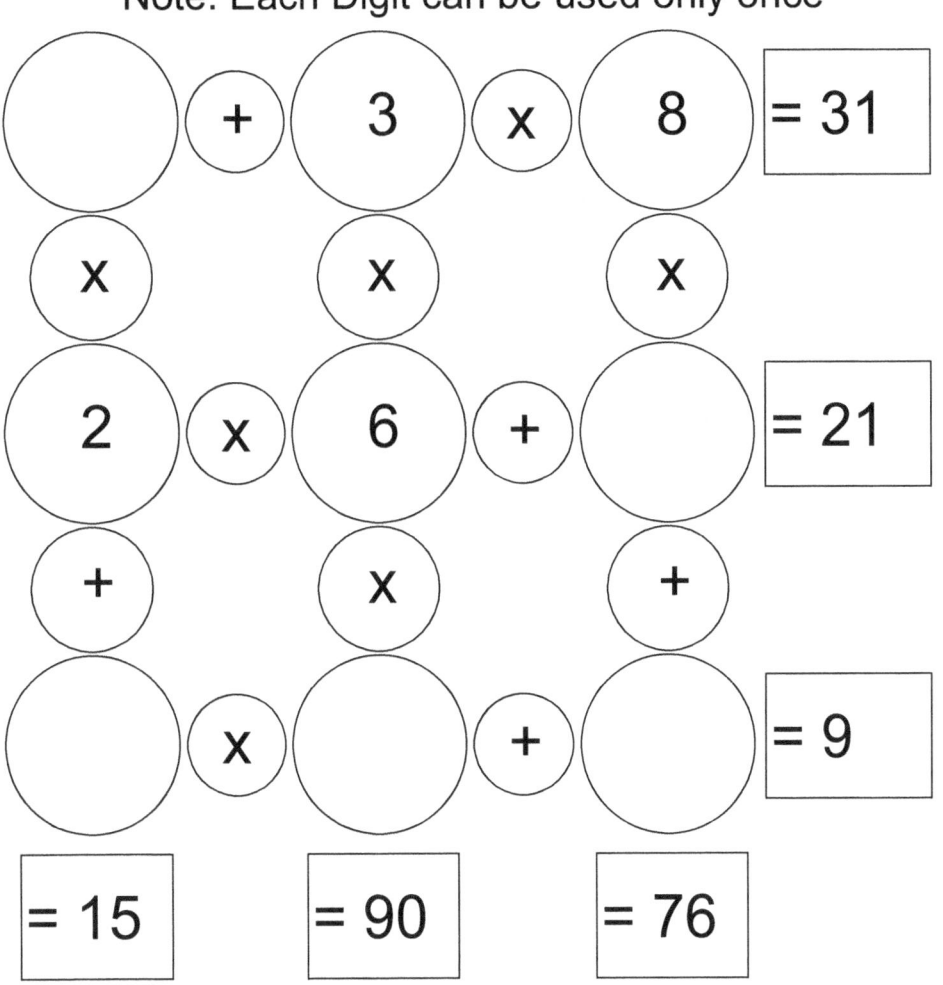

Solution at Page No 115

[27]

MATH IS FUN 5A

Puzzle Number 18

Digits to be used: 1,2,4,5,7

Note: Each Digit can be used only once

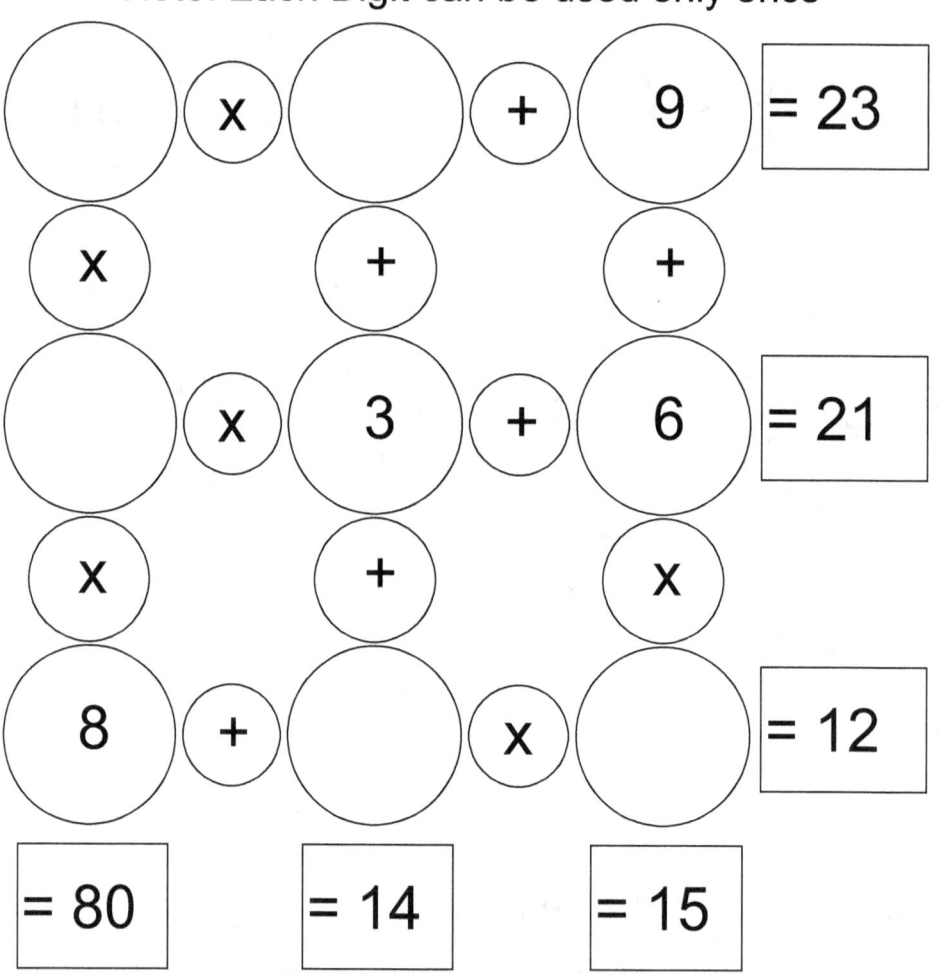

Solution at Page No 115

Puzzle Number 19

Digits to be used: 3,4,5,8,9

Note: Each Digit can be used only once

7 x 2 + ( ) = 18

x + +

1 x ( ) + ( ) = 14

+ + x

( ) + ( ) + 6 = 17

= 15   = 14   = 34

Solution at Page No 115

[29]

Puzzle Number 20

Digits to be used: 1,3,4,7,9

Note: Each Digit can be used only once

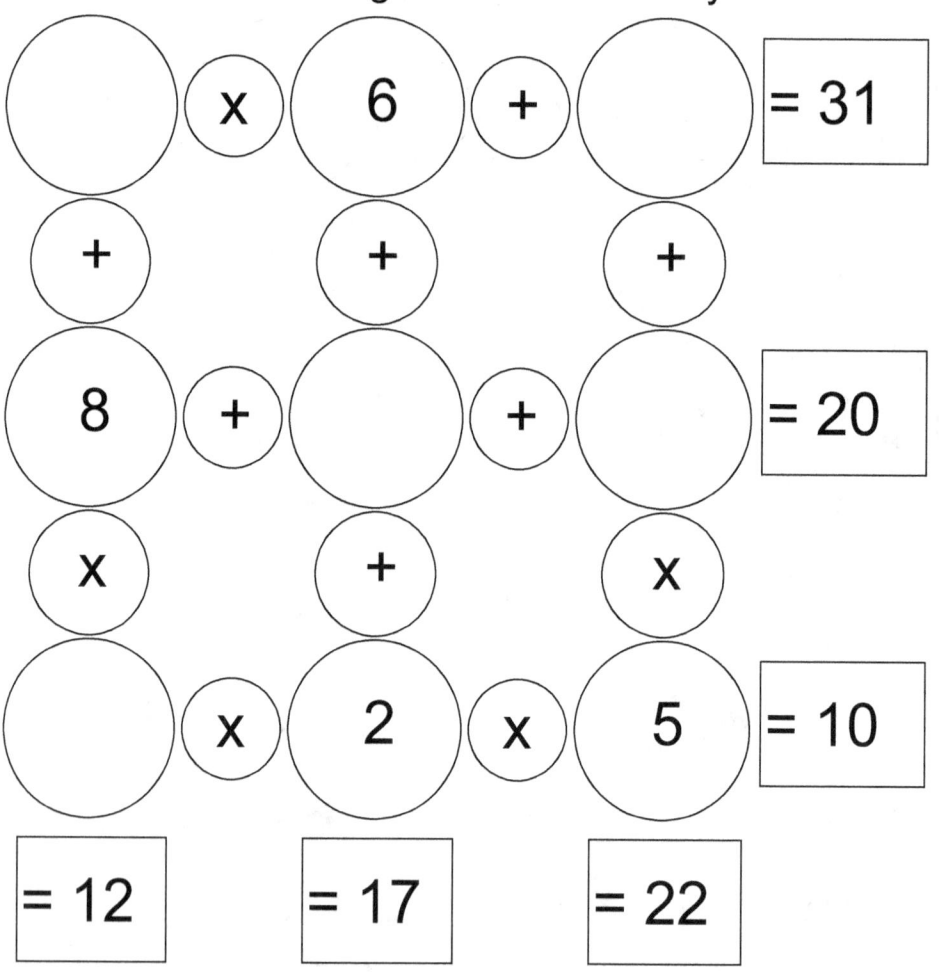

Solution at Page No 115

Puzzle Number 21

Digits to be used: 3,4,5,6,7

Note: Each Digit can be used only once

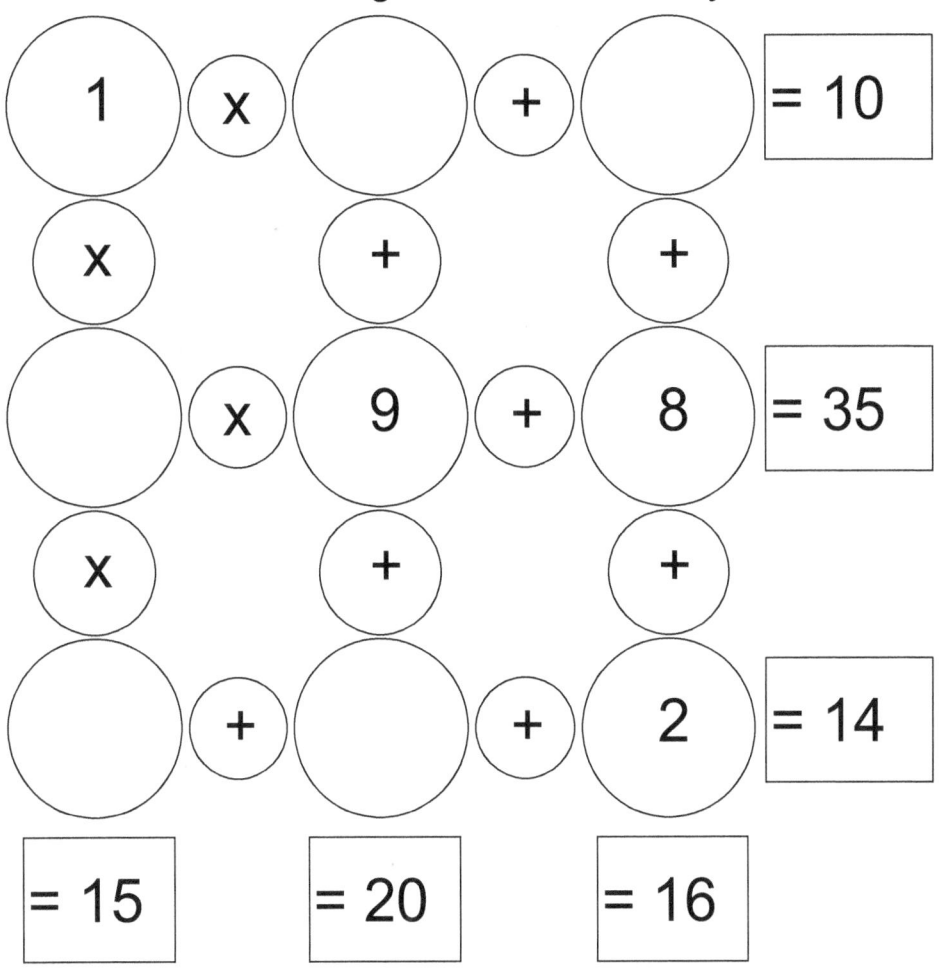

Solution at Page No 116

Puzzle Number 22

Digits to be used: 1,2,5,7,8

Note: Each Digit can be used only once

( ) + (4) x (3) = 20

+   x   x

( ) + ( ) x (9) = 68

+   x   x

(6) x ( ) x ( ) = 12

= 19   = 28   = 54

Solution at Page No 116

[32]

Puzzle Number 23

Digits to be used: 4,5,7,8,9

Note: Each Digit can be used only once

(6) + ( ) x (3) = 30

+   +   +

( ) + (2) x (1) = 7

+   +   +

( ) + ( ) x ( ) = 43

= 18   = 14   = 13

Solution at Page No 116

[33]

MATH IS FUN 5A

Puzzle Number 24

Digits to be used: 1,3,6,7,8

Note: Each Digit can be used only once

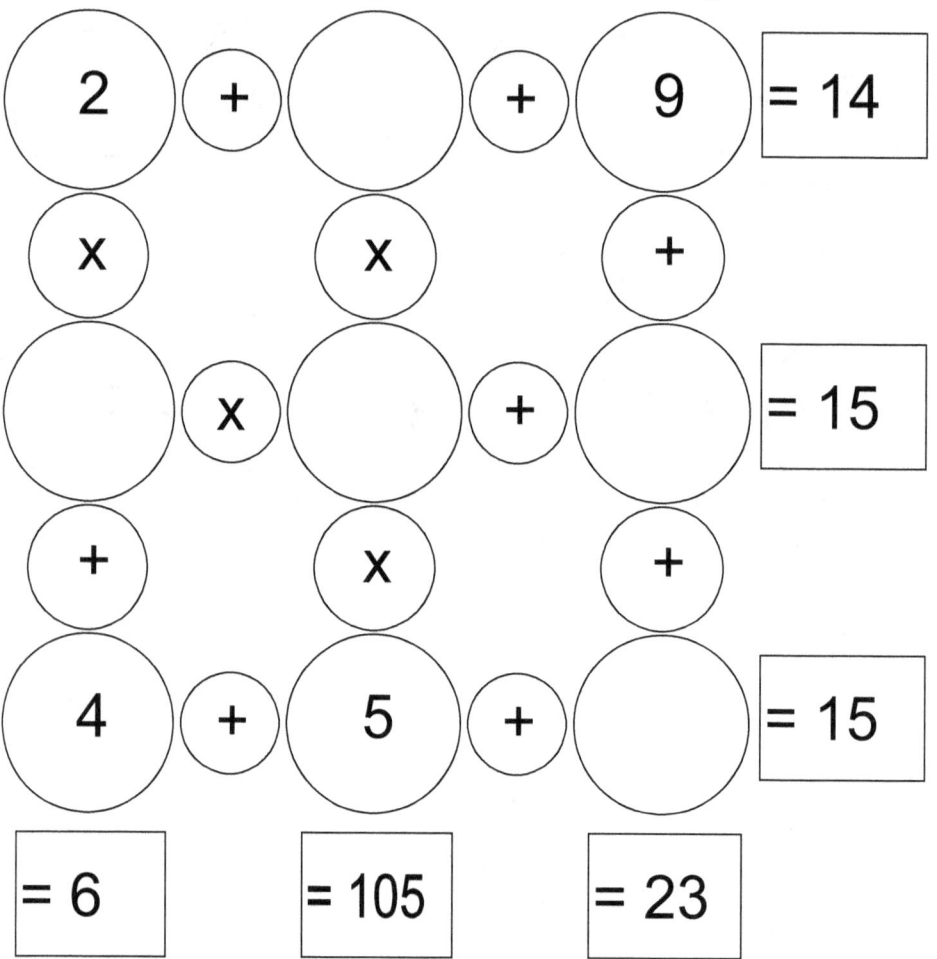

Solution at Page No 116

Puzzle Number 25

Digits to be used: 1,3,5,6,7

Note: Each Digit can be used only once

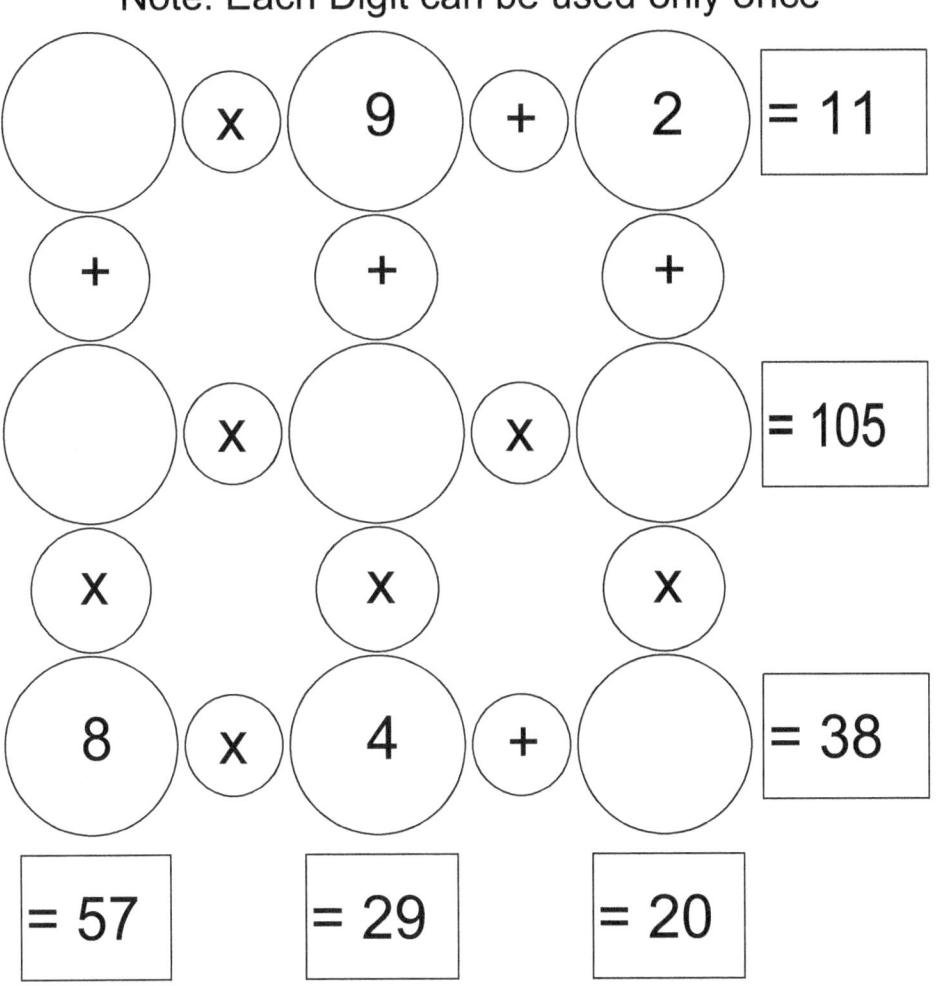

Solution at Page No 117

[35]

MATH IS FUN 5A

Puzzle Number 26

Digits to be used: 2,3,6,8,9

Note: Each Digit can be used only once

$$4 \times \bigcirc + 1 = 33$$
$$\times \quad + \quad \times$$
$$\bigcirc \times \bigcirc \times \bigcirc = 36$$
$$\times \quad \times \quad +$$
$$\bigcirc + 5 + 7 = 21$$
$$= 72 \quad = 38 \quad = 10$$

Solution at Page No 117

Puzzle Number 27

Digits to be used: 1,2,5,6,7

Note: Each Digit can be used only once

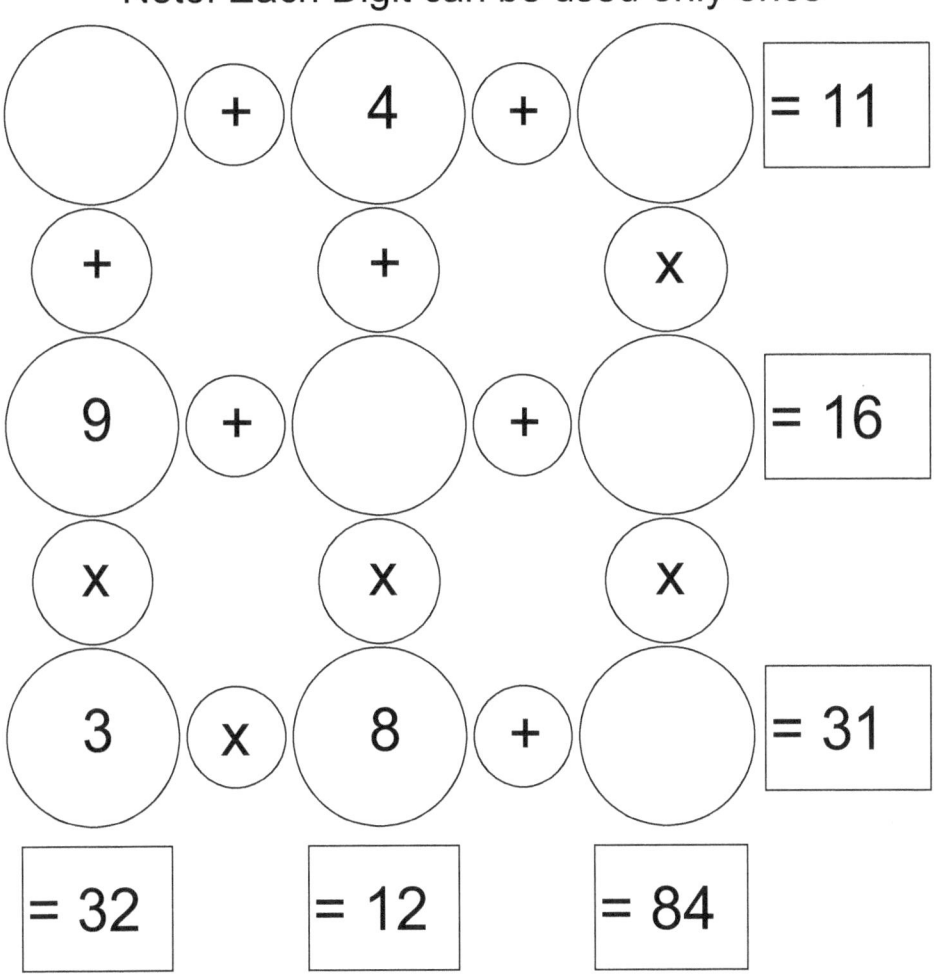

Solution at Page No 117

MATH IS FUN 5A

Puzzle Number 28

Digits to be used: 1,3,6,7,8

Note: Each Digit can be used only once

Solution at Page No 117

MATH IS FUN 5A

Puzzle Number 29

Digits to be used: 1,2,3,7,8

Note: Each Digit can be used only once

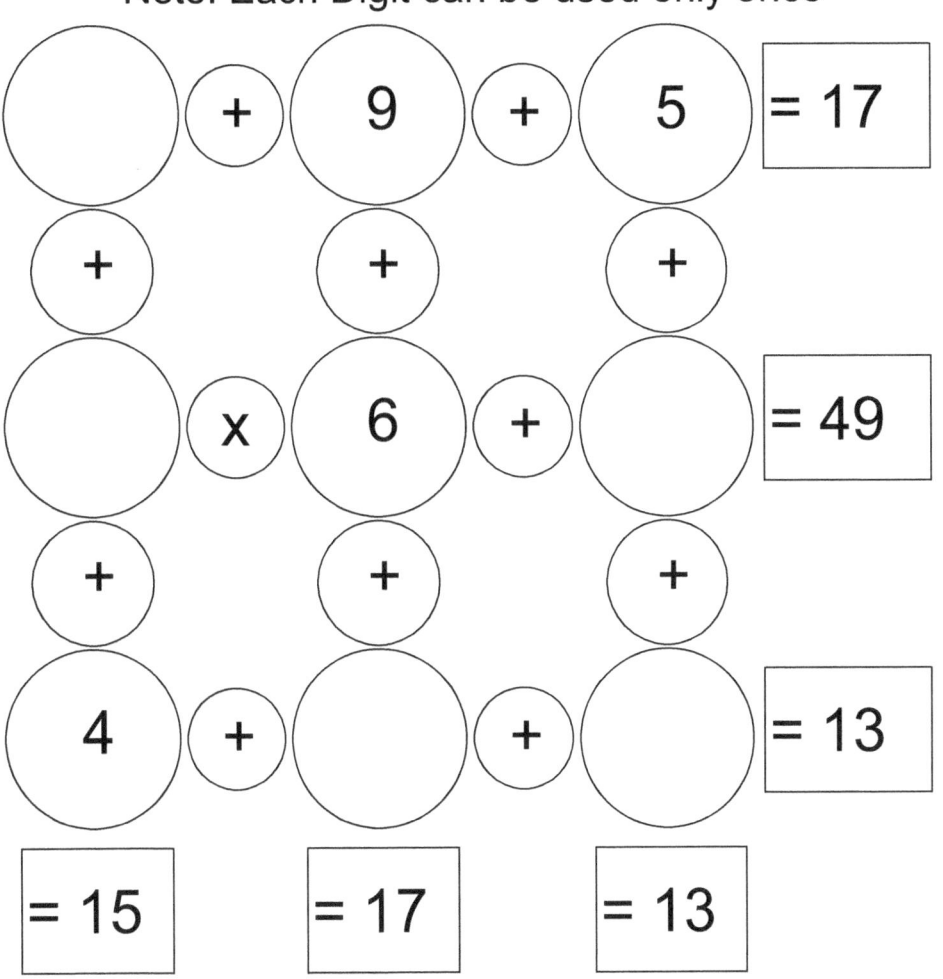

Solution at Page No 118

MATH IS FUN 5A

Puzzle Number 30

Digits to be used: 2,5,6,7,9

Note: Each Digit can be used only once

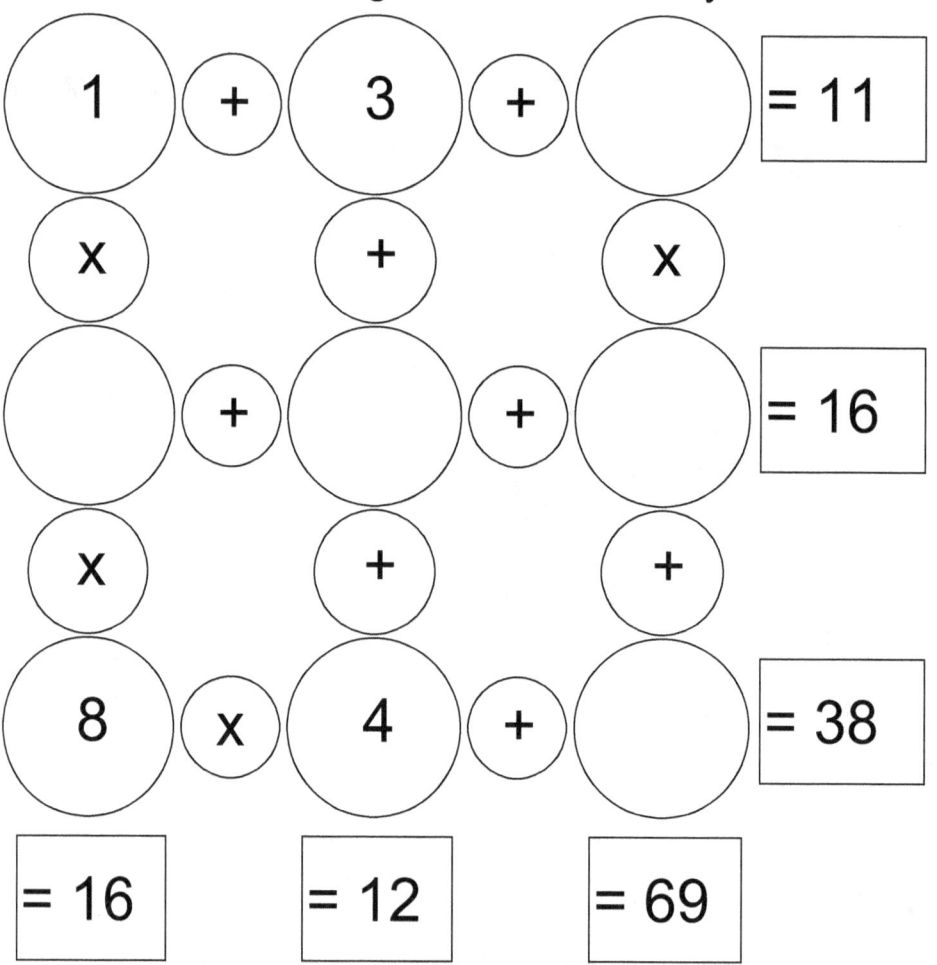

Solution at Page No 118

[40]

Puzzle Number 31

Digits to be used: 2,3,5,6,9

Note: Each Digit can be used only once

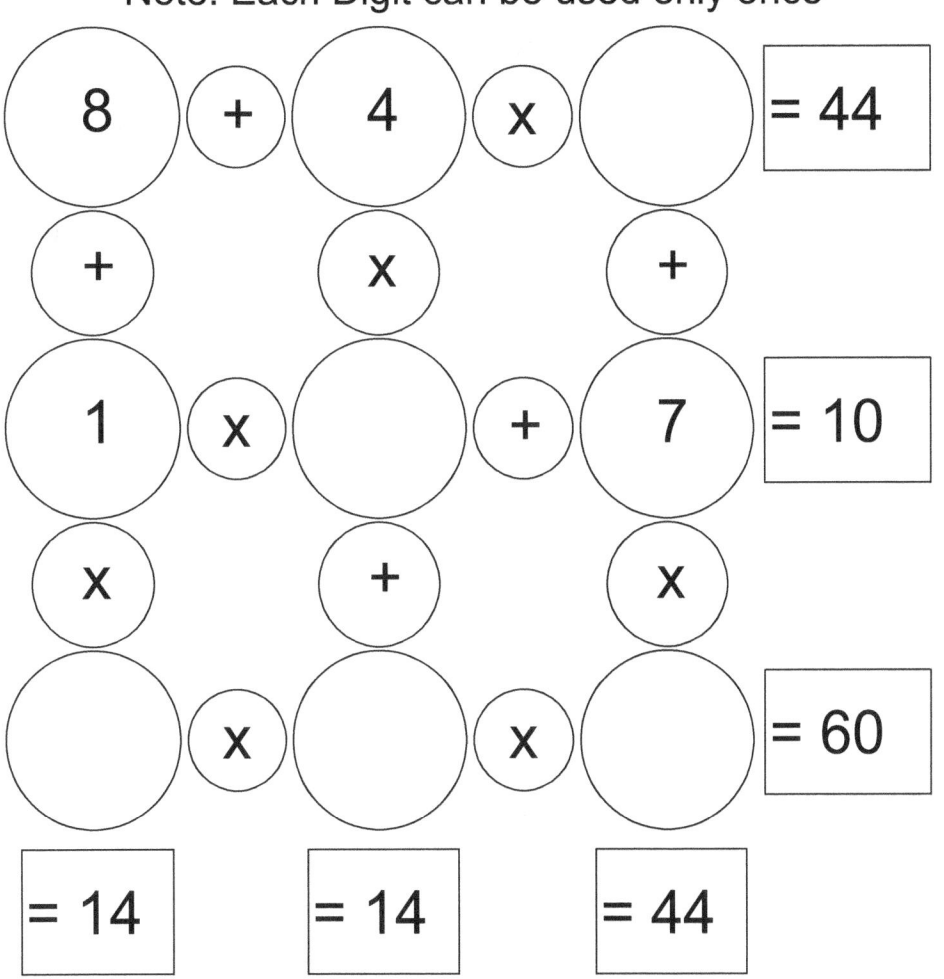

Solution at Page No 118

[41]

MATH IS FUN 5A

Puzzle Number 32

Digits to be used: 1,3,4,5,8

Note: Each Digit can be used only once

[ ] x 2 x [ ] = 24

x + +

[ ] + [ ] x 7 = 57

+ x +

[ ] x 9 + 6 = 51

= 9   = 74   = 16

Solution at Page No 118

MATH IS FUN 5A

Puzzle Number 33

Digits to be used: 1,2,3,7,8

Note: Each Digit can be used only once

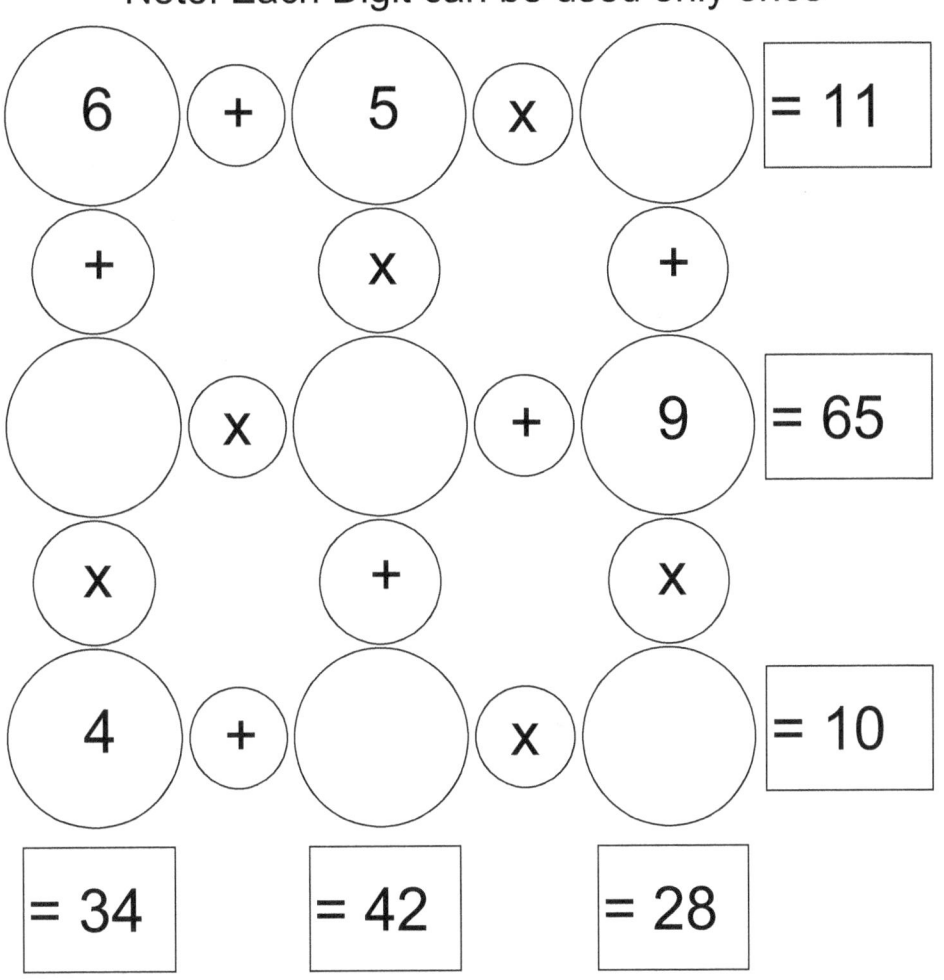

Solution at Page No 119

Puzzle Number 34

Digits to be used: 2,4,5,7,9

Note: Each Digit can be used only once

( ) + (3) x ( ) = 26

x   +   +

( ) + ( ) x (1) = 6

x   x   +

( ) + (8) x (6) = 57

= 90   = 35   = 14

Solution at Page No 119

[44]

MATH IS FUN 5A

Puzzle Number 35

Digits to be used: 2,4,5,7,9

Note: Each Digit can be used only once

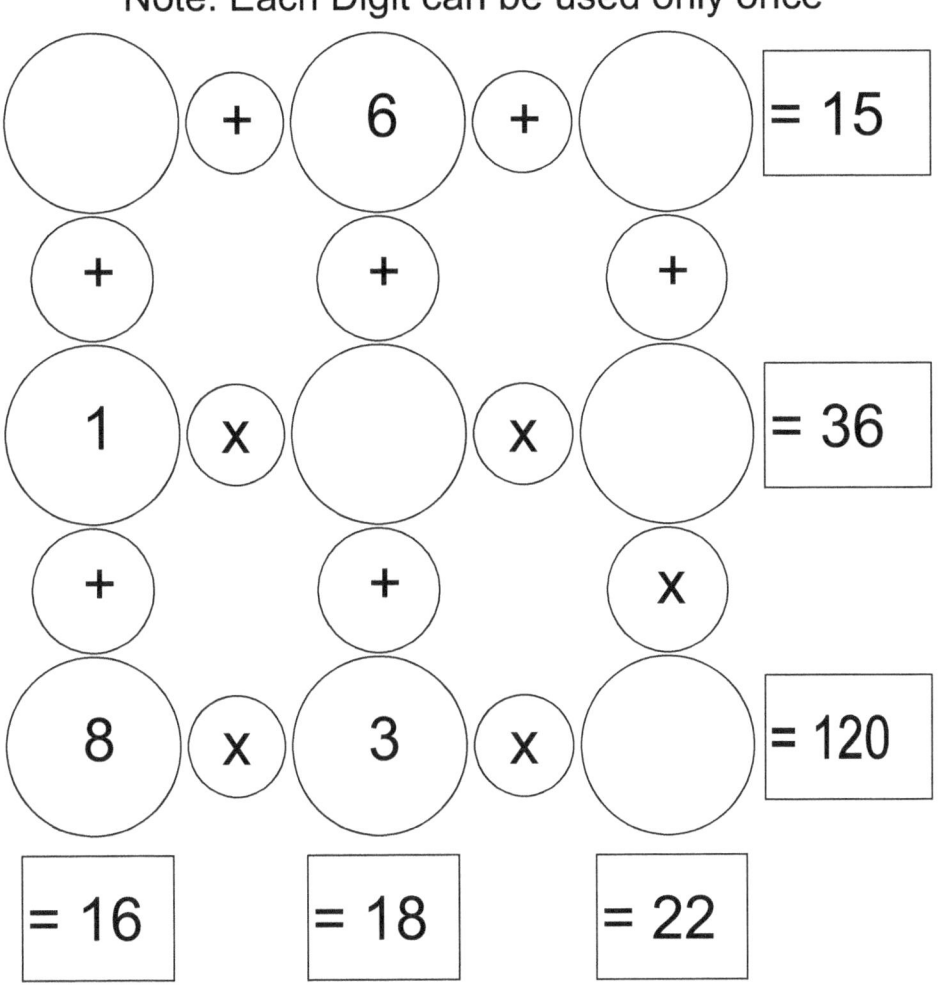

Solution at Page No 119

Puzzle Number 36

Digits to be used: 1,2,3,6,8

Note: Each Digit can be used only once

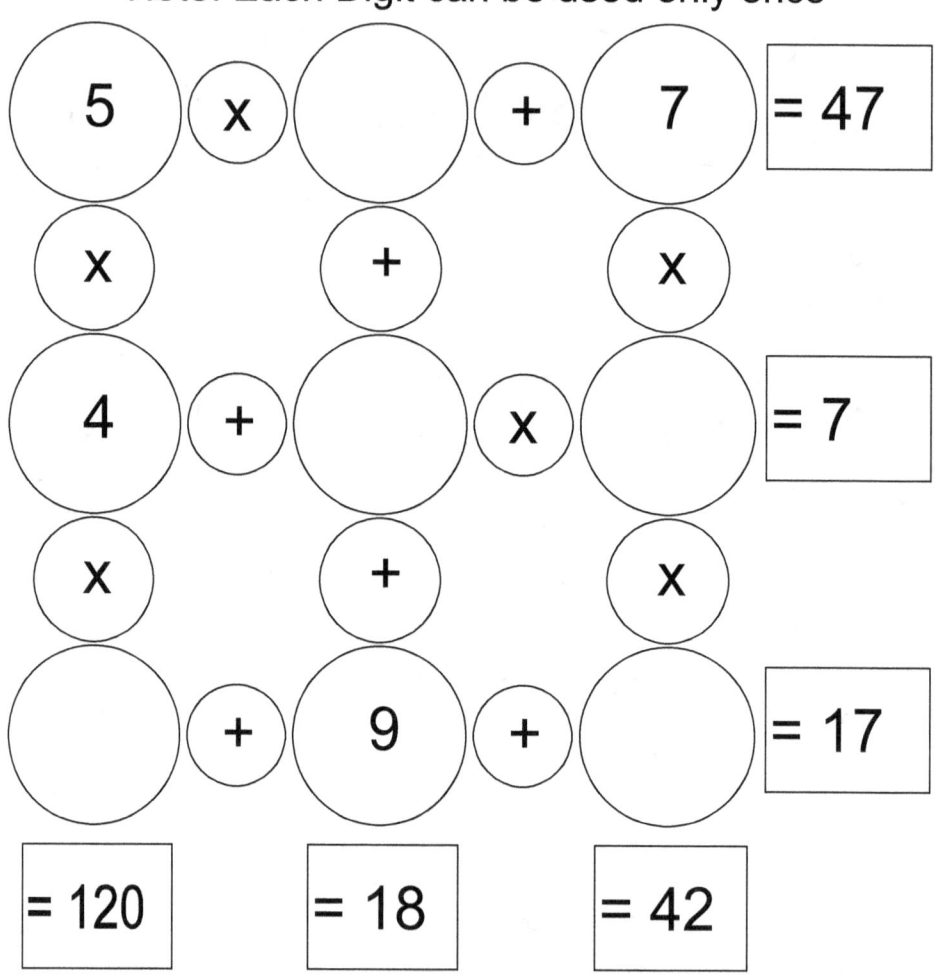

Solution at Page No 119

[46]

MATH IS FUN 5A

Puzzle Number 37

Digits to be used: 2,3,4,5,7

Note: Each Digit can be used only once

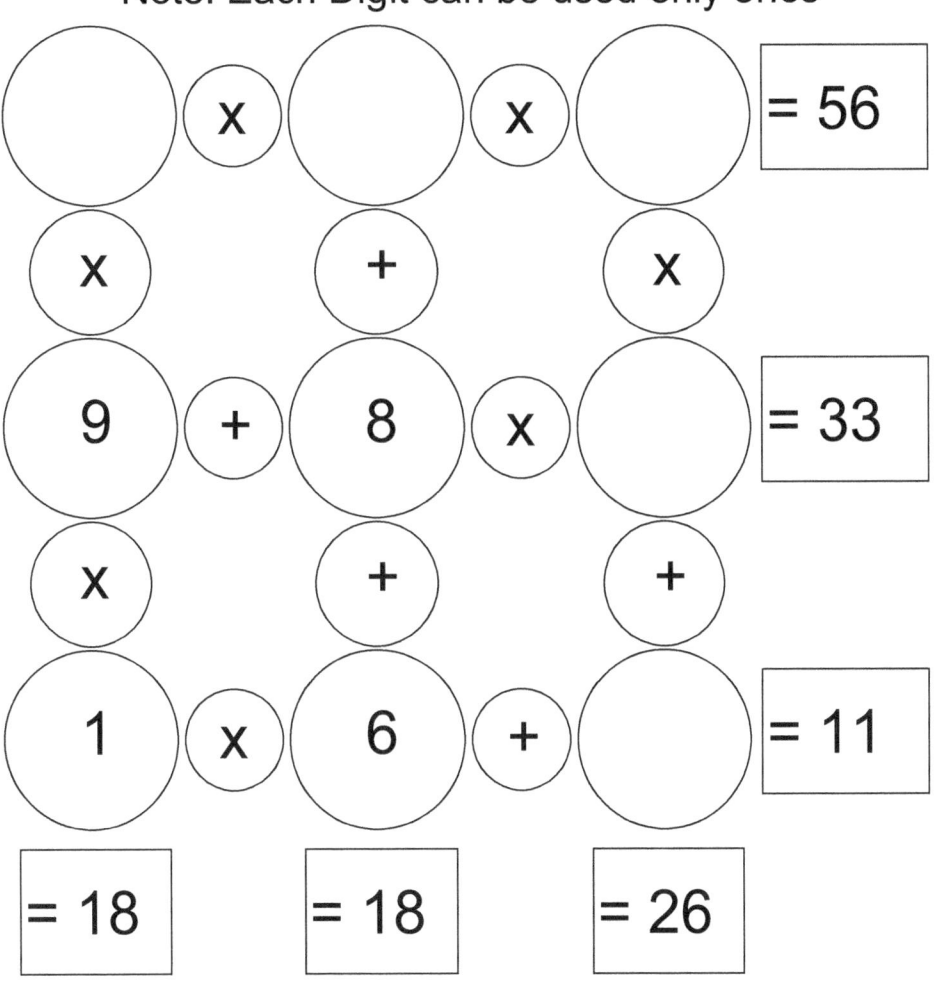

Solution at Page No 120

MATH IS FUN 5A

Puzzle Number 38

Digits to be used: 2,3,6,7,8

Note: Each Digit can be used only once

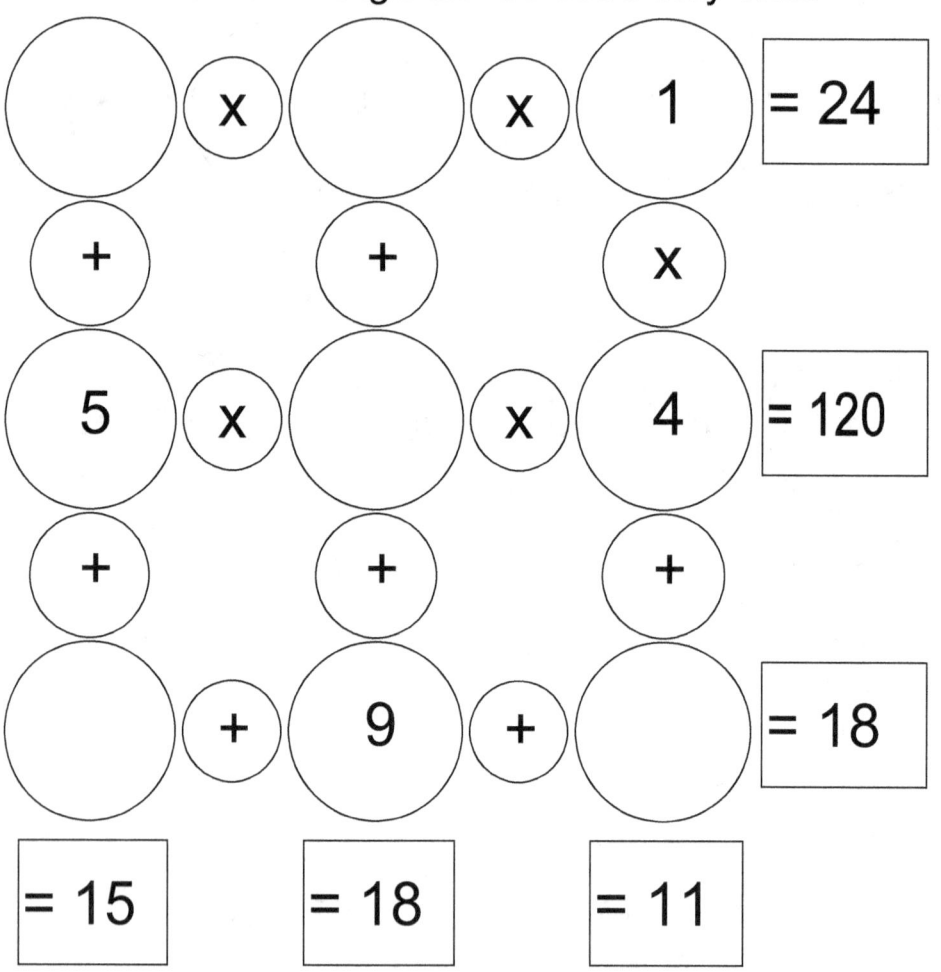

Solution at Page No 120

[48]

Puzzle Number 39

Digits to be used: 1,2,4,6,9

Note: Each Digit can be used only once

◯	+	◯	+	◯	= 14
+		+		+	
5	+	◯	X	7	= 19
X		X		+	
8	+	◯	X	3	= 26
= 41		= 16		= 19	

Solution at Page No 120

Puzzle Number 40

Digits to be used: 1,2,4,7,9

Note: Each Digit can be used only once

5	X	( )	+ ( ) = 22

(5) X ( ) + ( ) = 22

X   +   +

(6) X ( ) + (8) = 50

+   +   +

( ) + (3) X ( ) = 28

= 31   = 14   = 19

Solution at Page No 120

MATH IS FUN 5A

Puzzle Number 41

Digits to be used: 1,3,5,6,7

Note: Each Digit can be used only once

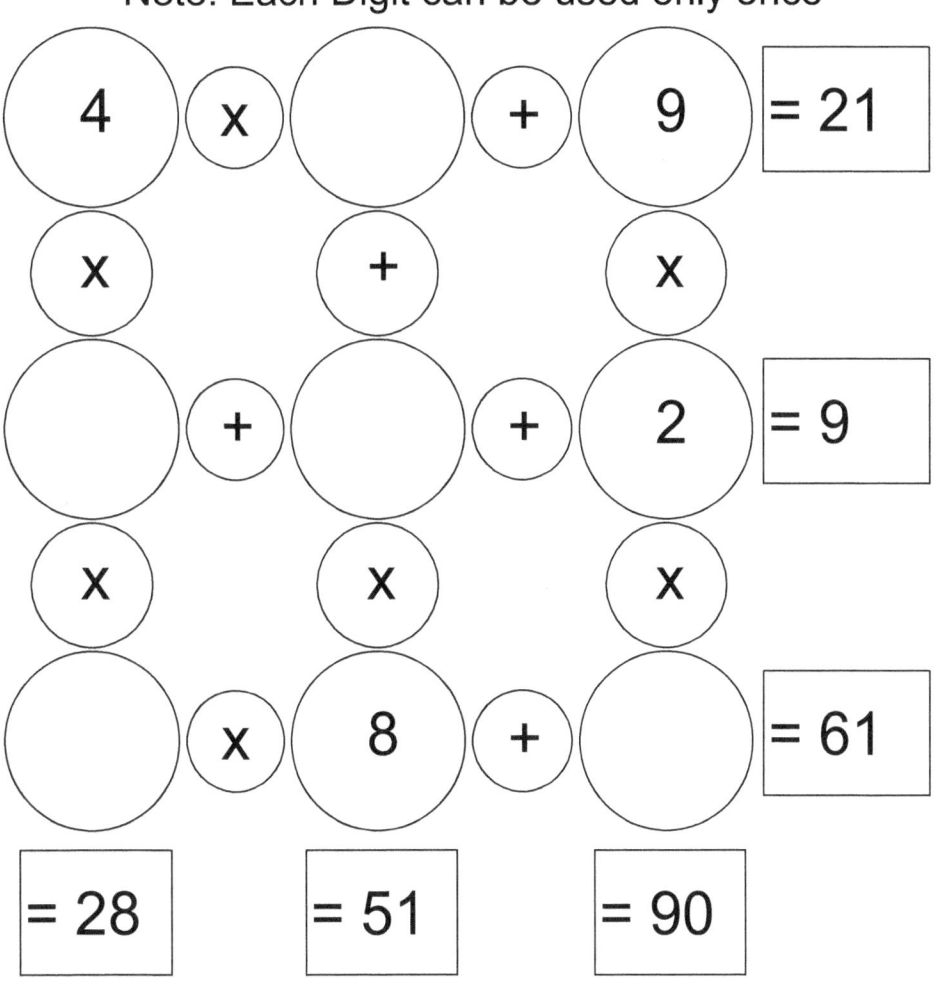

Solution at Page No 121

Puzzle Number 42

Digits to be used: 1,5,6,8,9

Note: Each Digit can be used only once

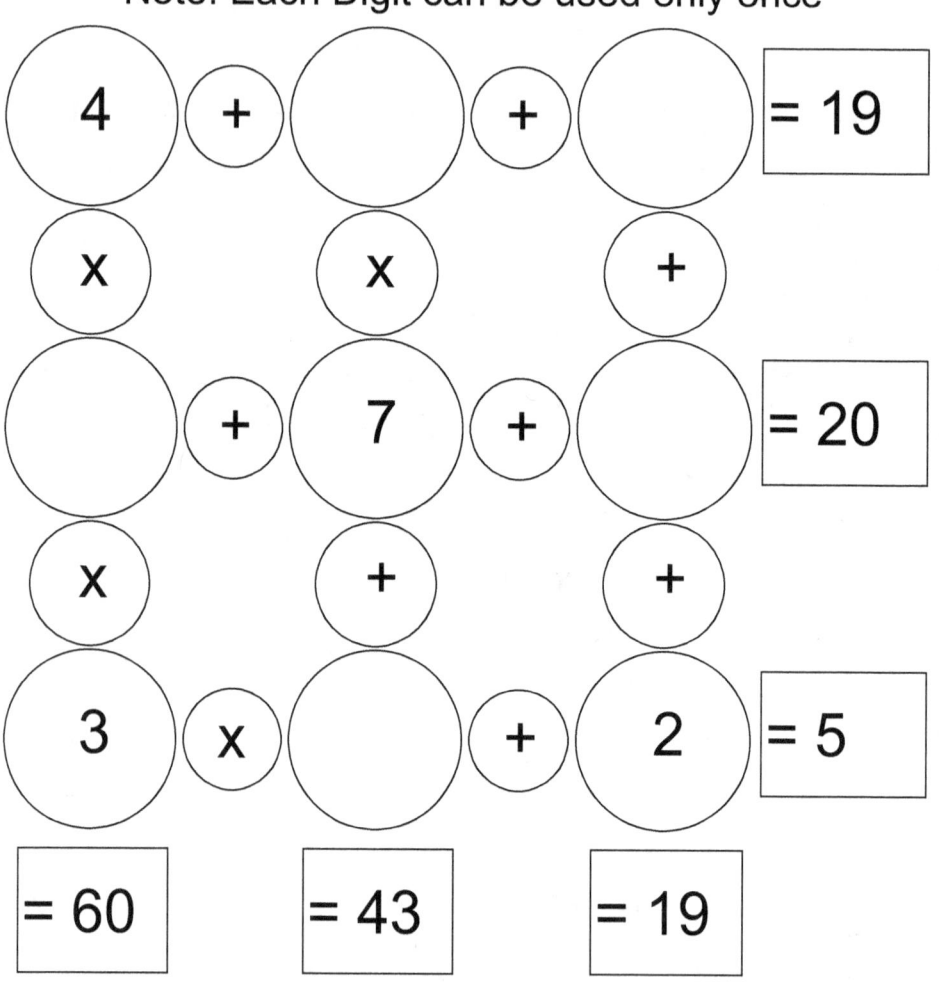

Solution at Page No 121

Puzzle Number 43

Digits to be used: 1,3,5,6,7

Note: Each Digit can be used only once

4 × ___ + ___ = 23

+   +   ×

___ + 2 + 9 = 18

×   ×   +

___ + 8 + ___ = 15

= 46   = 21   = 28

Solution at Page No 121

[53]

MATH IS FUN 5A

Puzzle Number 44

Digits to be used: 1,2,3,4,6

Note: Each Digit can be used only once

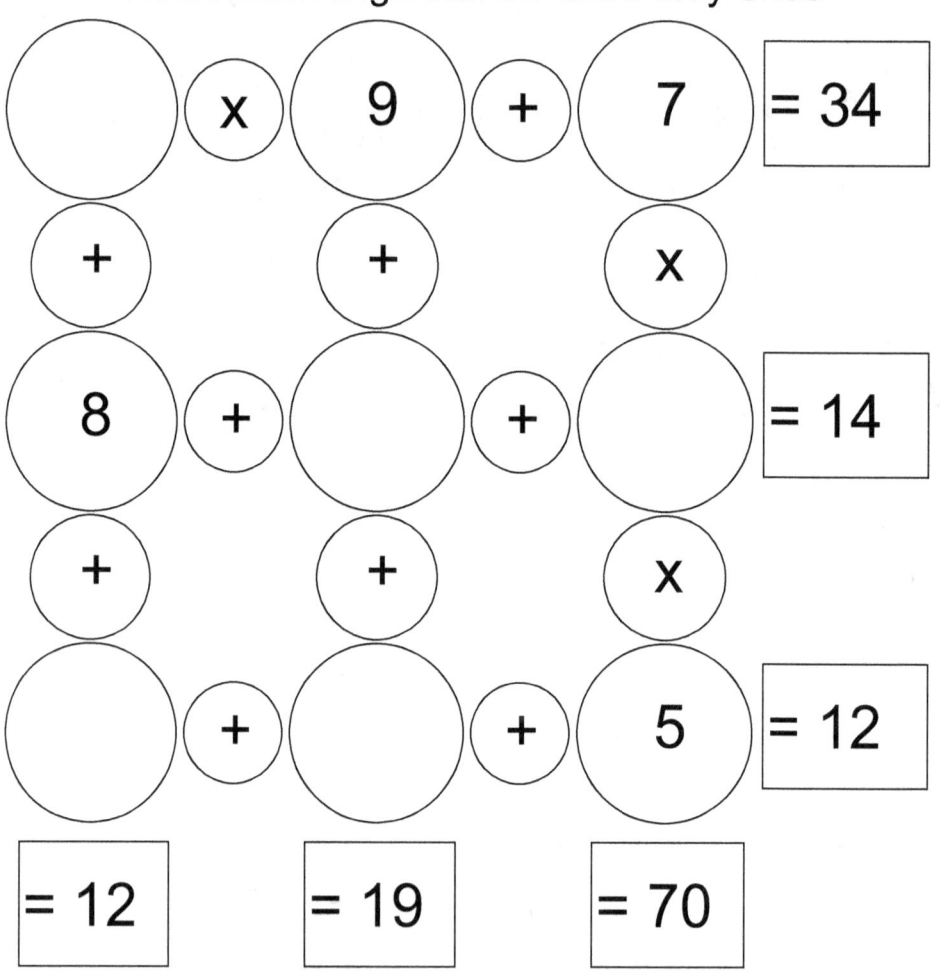

Solution at Page No 121

Puzzle Number 45

Digits to be used: 1,3,6,8,9

Note: Each Digit can be used only once

$$(2) + (\quad) \times (7) = 58$$
$$\times \quad \times \quad +$$
$$(\quad) + (5) + (\quad) = 9$$
$$+ \quad + \quad +$$
$$(\quad) \times (\quad) + (4) = 58$$
$$= 8 \qquad = 49 \qquad = 14$$

Solution at Page No 122

Puzzle Number 46

Digits to be used: 2,4,6,8,9

Note: Each Digit can be used only once

$$7 + (\quad) \times 5 = 37$$
$$+ \qquad + \qquad +$$
$$(\quad) \times (\quad) \times 1 = 16$$
$$\times \qquad \times \qquad +$$
$$(\quad) \times 3 \times (\quad) = 108$$
$$= 25 \qquad = 30 \qquad = 10$$

Solution at Page No 122

Puzzle Number 47

Digits to be used: 1,3,4,8,9

Note: Each Digit can be used only once

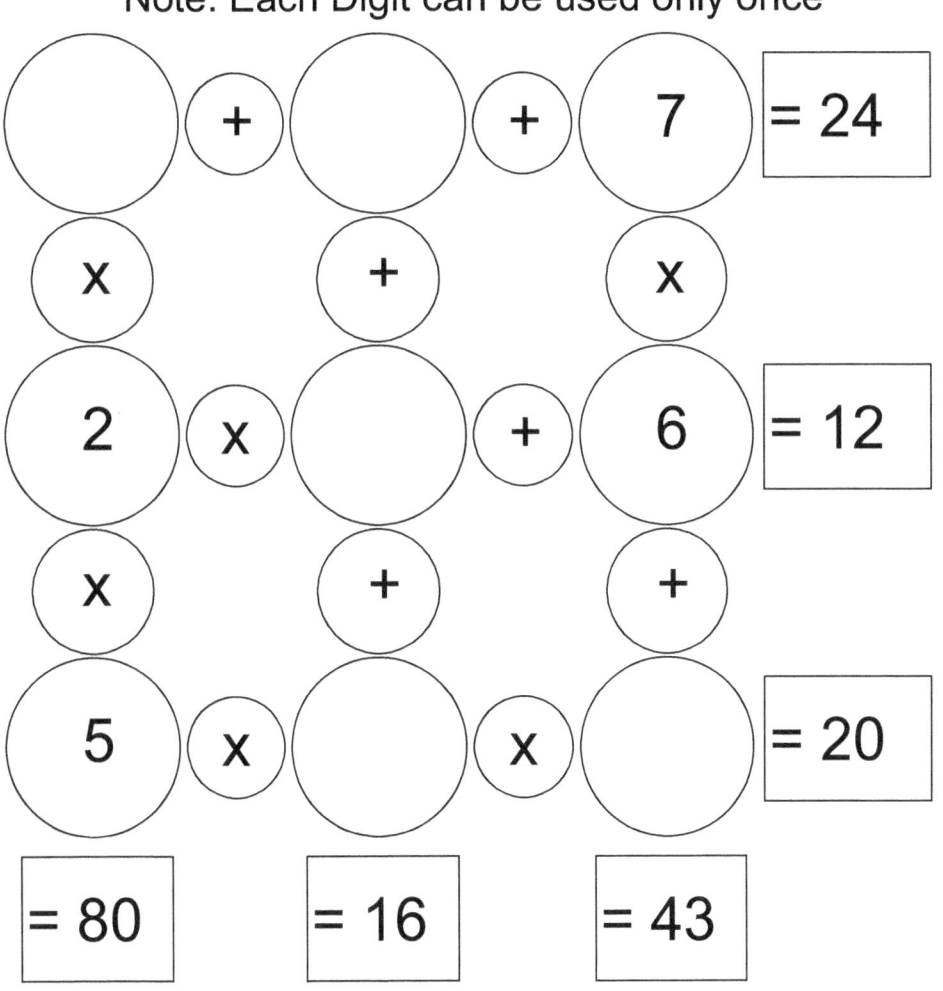

Solution at Page No 122

Puzzle Number 48

Digits to be used: 1,3,4,5,7

Note: Each Digit can be used only once

6 + 9 x ( ) = 69

x   x   +

2 x ( ) x ( ) = 24

x   +   x

( ) x 8 + ( ) = 13

= 12   = 35   = 27

Solution at Page No 122

[58]

MATH IS FUN 5A

Puzzle Number 49

Digits to be used: 1,2,3,5,7

Note: Each Digit can be used only once

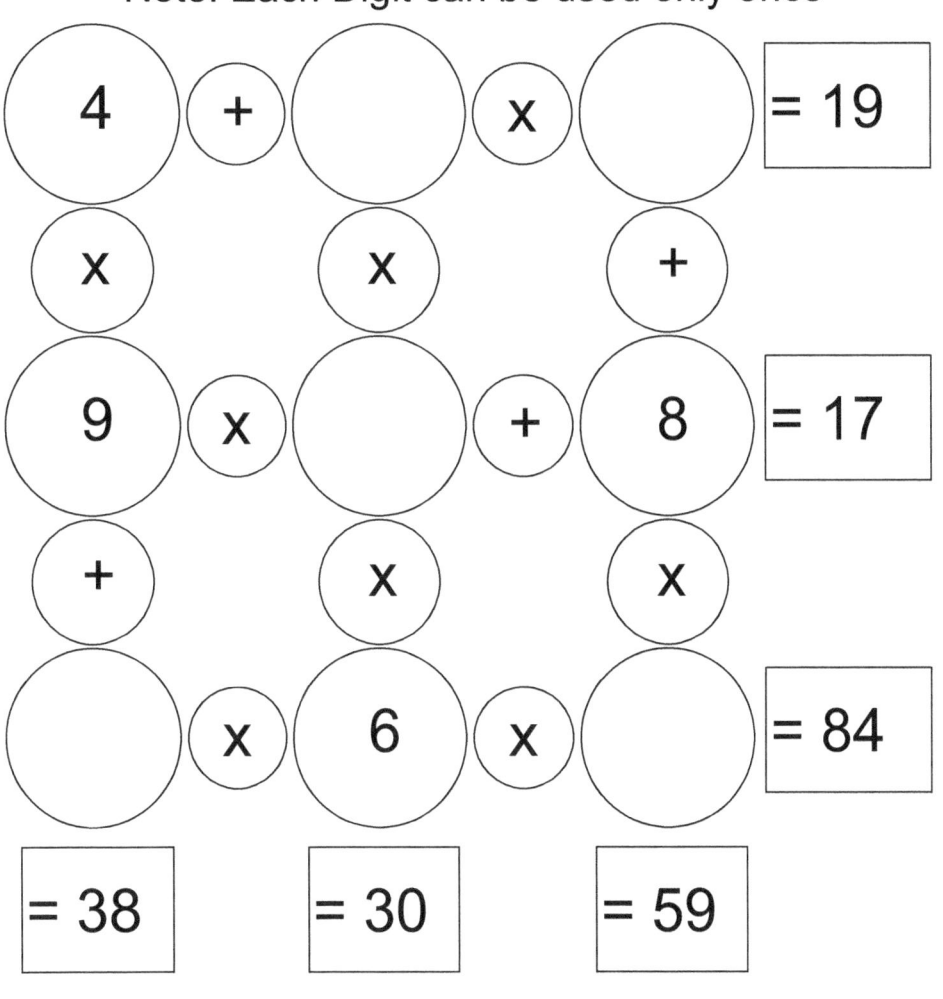

Solution at Page No 123

MATH IS FUN 5A

Puzzle Number 50

Digits to be used: 2,5,6,7,9

Note: Each Digit can be used only once

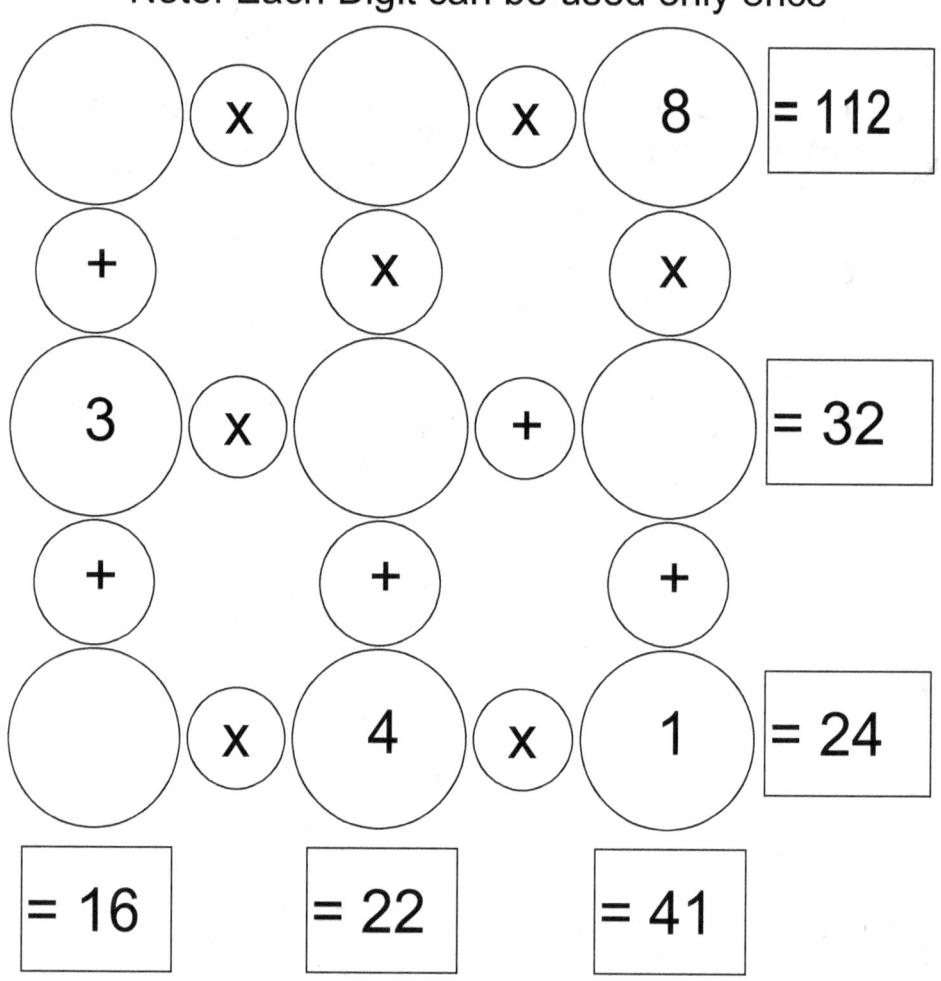

Solution at Page No 123

Puzzle Number 51

Digits to be used: 4,6,7,8,9

Note: Each Digit can be used only once

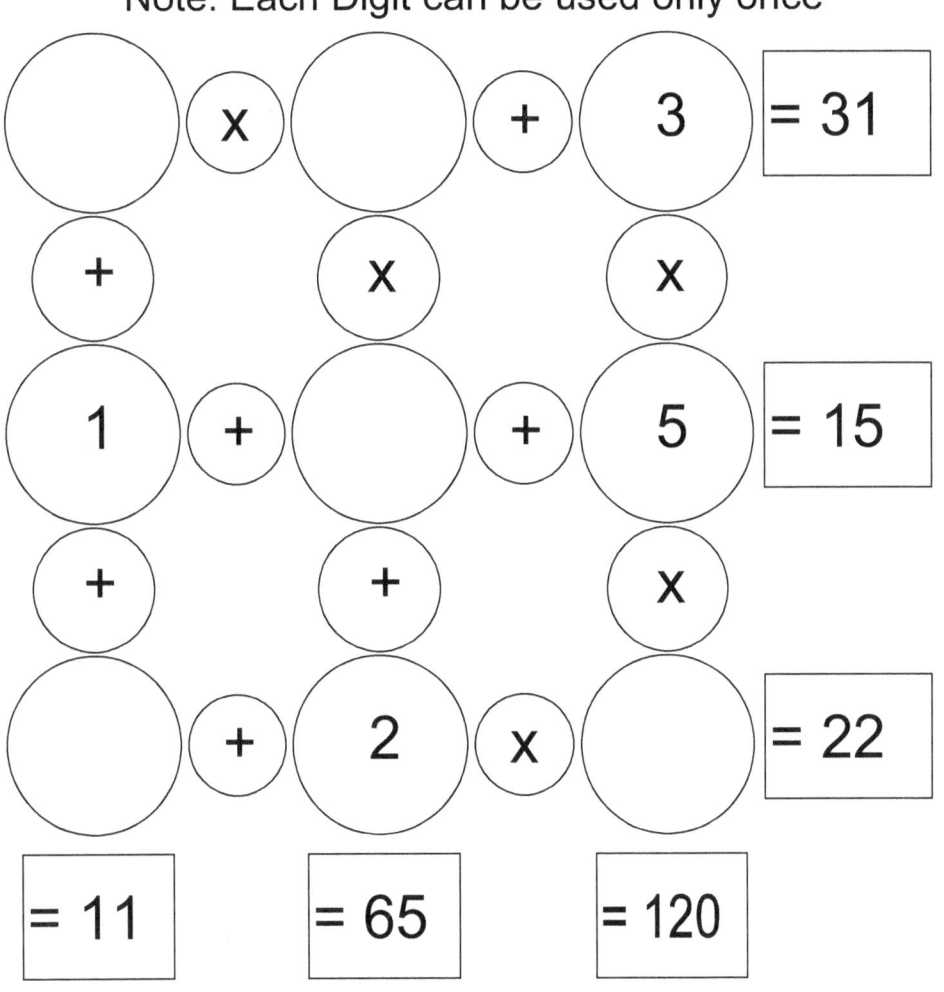

Solution at Page No 123

Puzzle Number 52

Digits to be used: 1,2,4,6,9

Note: Each Digit can be used only once

( ) + (8) + (7) = 16

+ + +

(5) x (3) + ( ) = 19

+ x +

( ) x ( ) + ( ) = 56

= 12   = 35   = 13

Solution at Page No 123

Puzzle Number 53

Digits to be used: 1,3,4,6,7

Note: Each Digit can be used only once

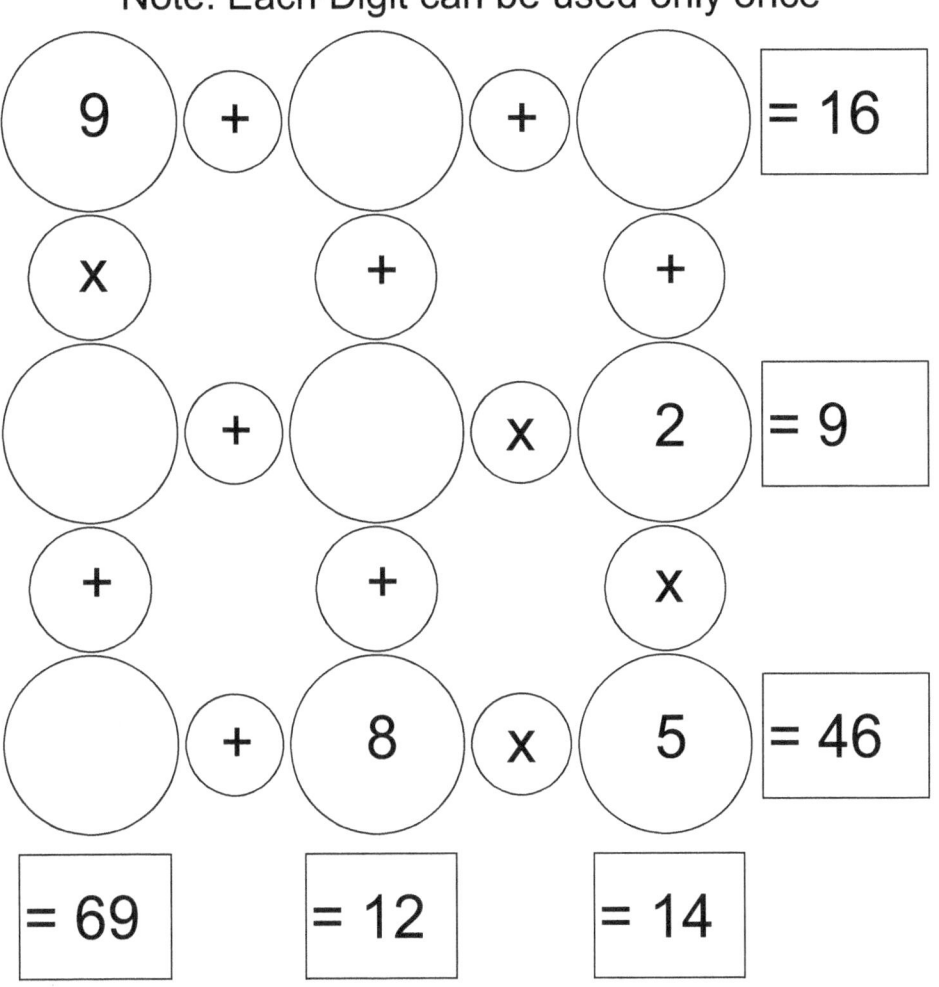

Solution at Page No 124

[63]

MATH IS FUN 5A

Puzzle Number 54

Digits to be used: 2,3,4,6,9

Note: Each Digit can be used only once

Solution at Page No 124

[64]

MATH IS FUN 5A

Puzzle Number 55

Digits to be used: 1,2,5,8,9

Note: Each Digit can be used only once

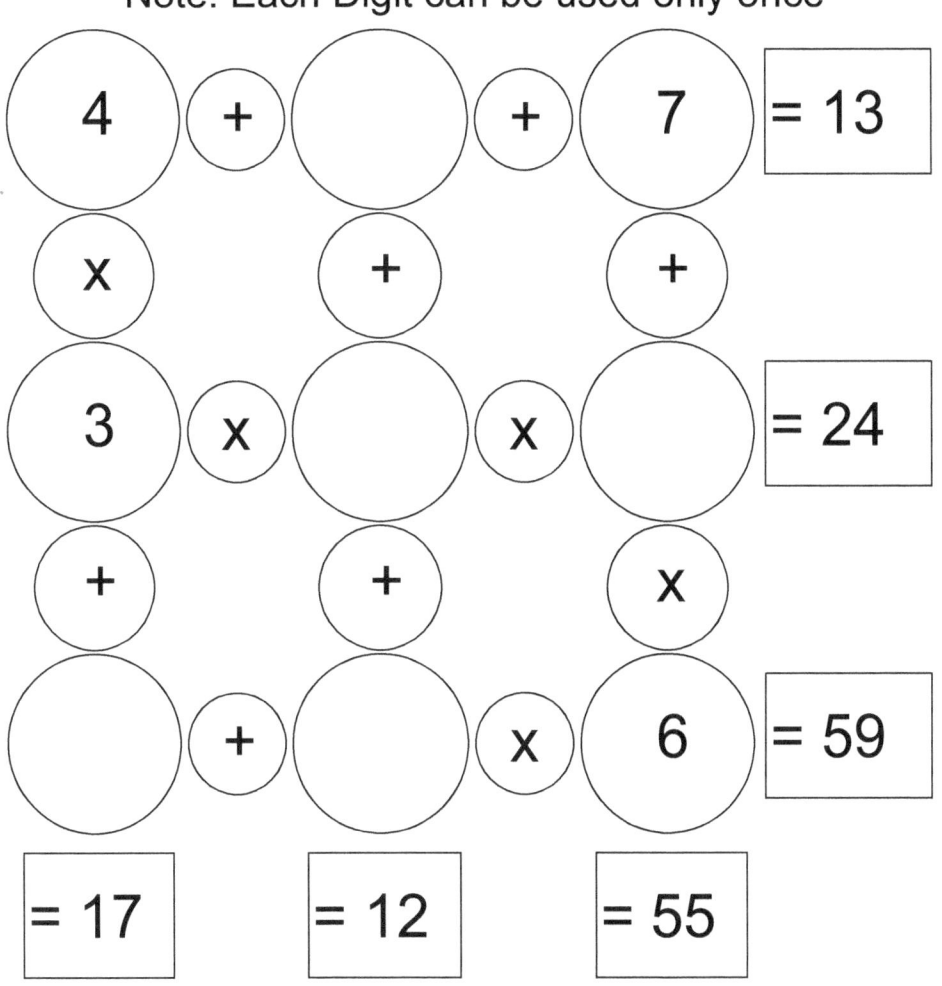

Solution at Page No 124

Puzzle Number 56

Digits to be used: 1,2,5,7,8

Note: Each Digit can be used only once

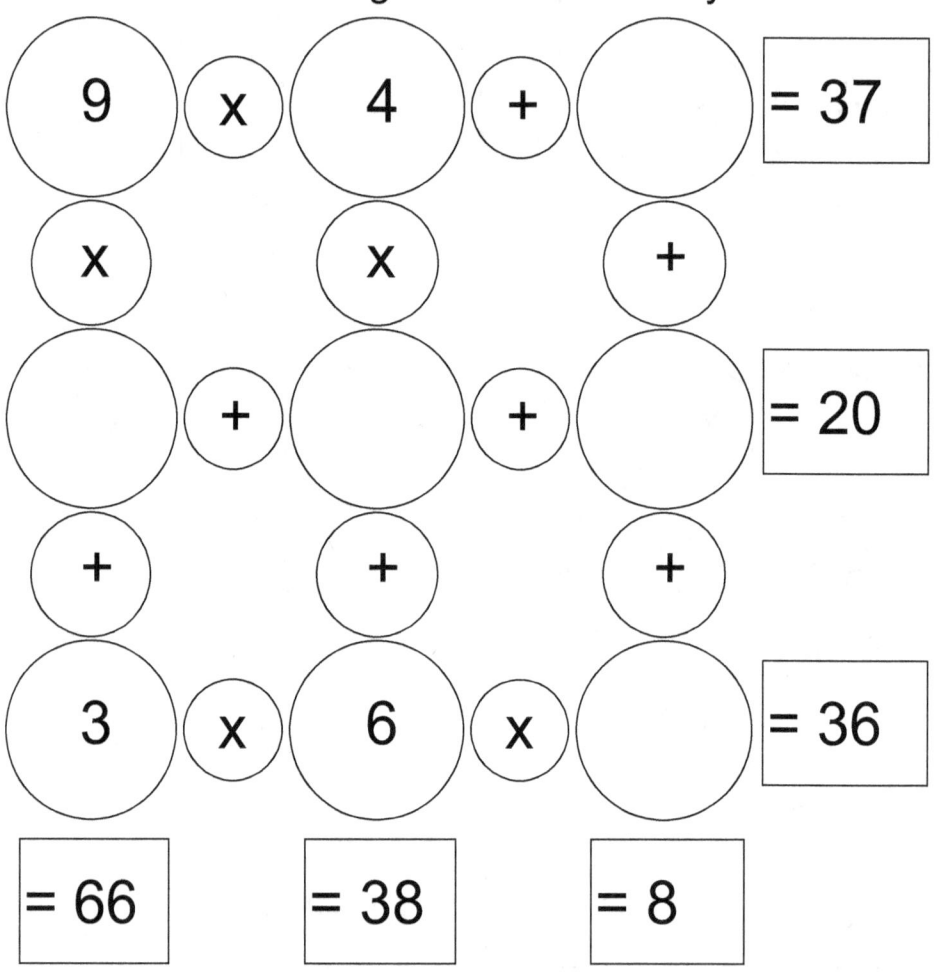

Solution at Page No 124

Puzzle Number 57

Digits to be used: 2,4,6,7,9

Note: Each Digit can be used only once

(8) + ( ) + ( )	= 17	

(8) + ( ) + ( ) = 17

+   ×   ×

( ) + (1) + ( ) = 16

+   ×   ×

( ) + (5) × (3) = 19

= 21   = 35   = 36

Solution at Page No 125

MATH IS FUN 5A

Puzzle Number 58

Digits to be used: 1,4,5,7,9

Note: Each Digit can be used only once

6 × 3 + ( )	= 22	
×	+	+
( ) + 2 × ( )	= 9	
+	+	+
8 + ( ) + ( )	= 22	
= 50	= 10	= 14

Solution at Page No 125

[68]

Puzzle Number 59

Digits to be used: 3,5,6,8,9

Note: Each Digit can be used only once

$$1 \times 4 \times \bigcirc = 24$$

$$\times \qquad + \qquad \times$$

$$\bigcirc + \bigcirc \times \bigcirc = 33$$

$$+ \qquad \times \qquad +$$

$$\bigcirc + 7 \times 2 = 19$$

$$= 14 \qquad = 60 \qquad = 20$$

Solution at Page No 125

MATH IS FUN 5A

Puzzle Number 60

Digits to be used: 3,5,6,8,9

Note: Each Digit can be used only once

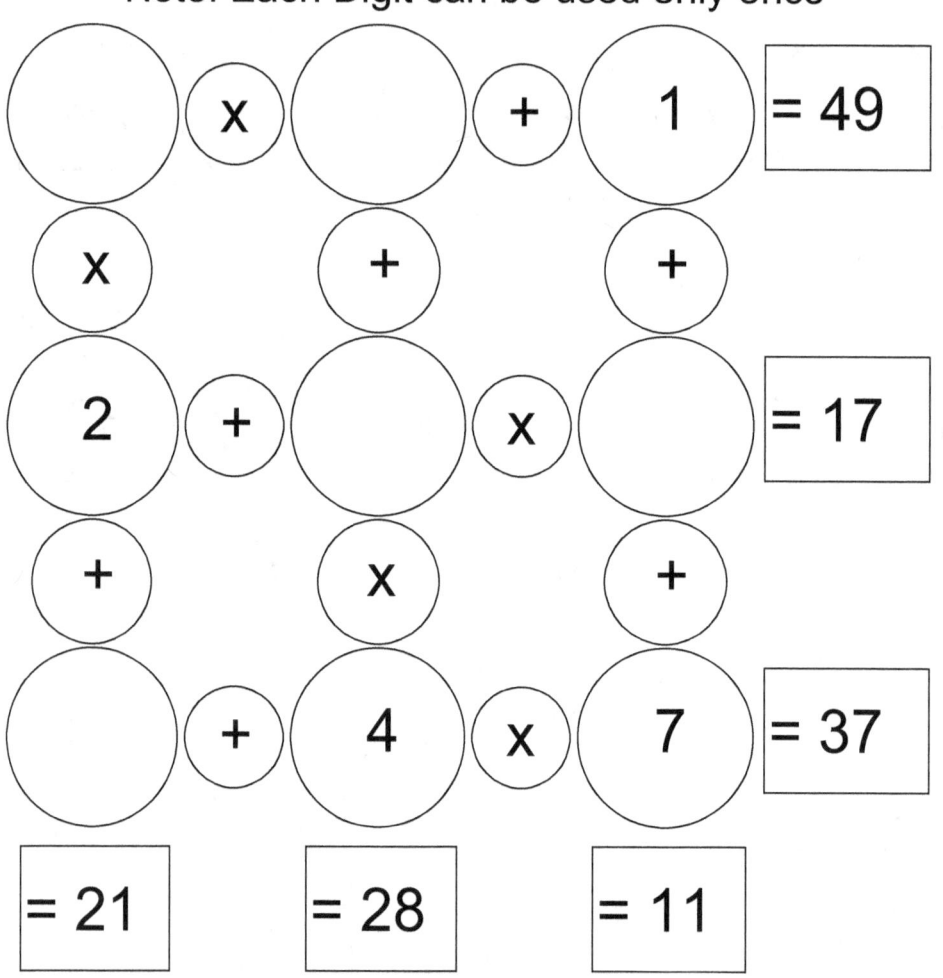

Solution at Page No 125

Puzzle Number 61

Digits to be used: 1,2,4,6,7

Note: Each Digit can be used only once

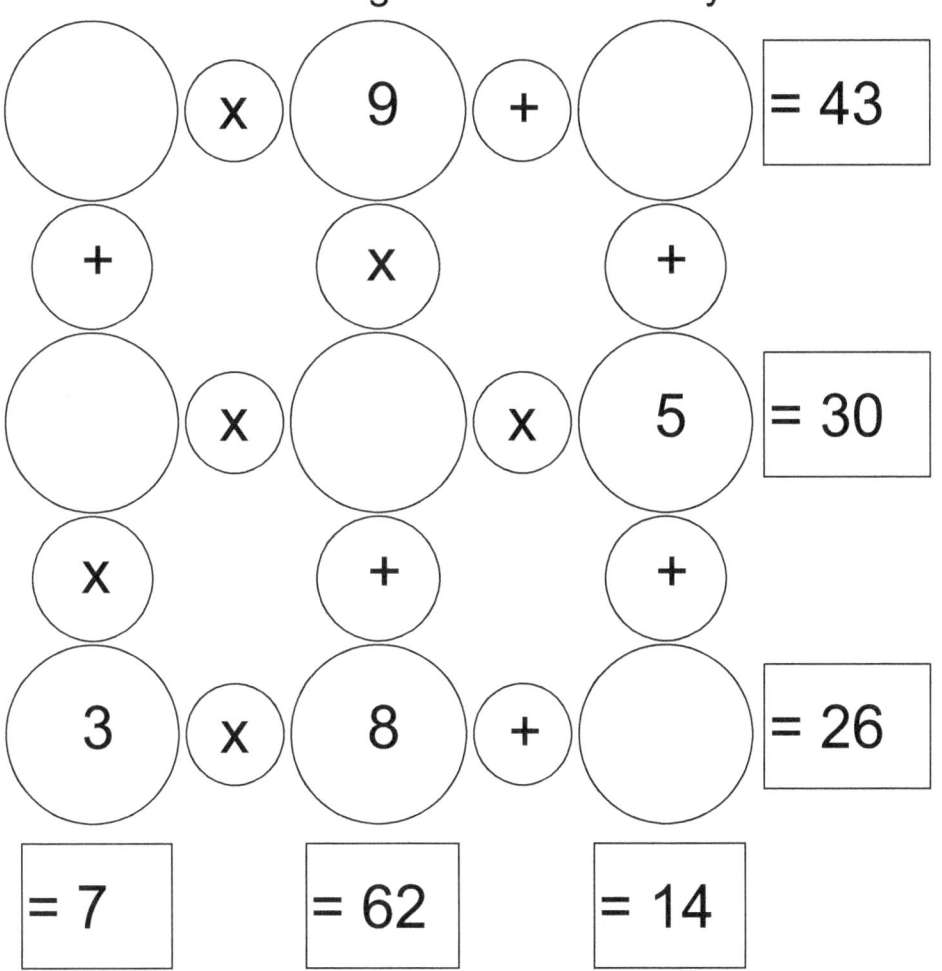

Solution at Page No 126

Puzzle Number 62

Digits to be used: 1,3,4,5,8

Note: Each Digit can be used only once

7 x ___ + ___ = 36

x + x

___ x 9 + 2 = 47

x x +

___ + ___ + 6 = 10

= 105   = 13   = 22

Solution at Page No 126

MATH IS FUN 5A

Puzzle Number 63

Digits to be used: 1,2,3,6,7

Note: Each Digit can be used only once

4	x	◯	+	8	= 20

(4) x (◯) + (8) = 20

(4) + ... (◯) x ... (8) +

(5) + (◯) x (9) = 59

(5) + ... (◯) + ... (9) +

(◯) + (◯) + (◯) = 10

= 11   = 25   = 18

Solution at Page No 126

Puzzle Number 64

Digits to be used: 2,4,5,6,9

Note: Each Digit can be used only once

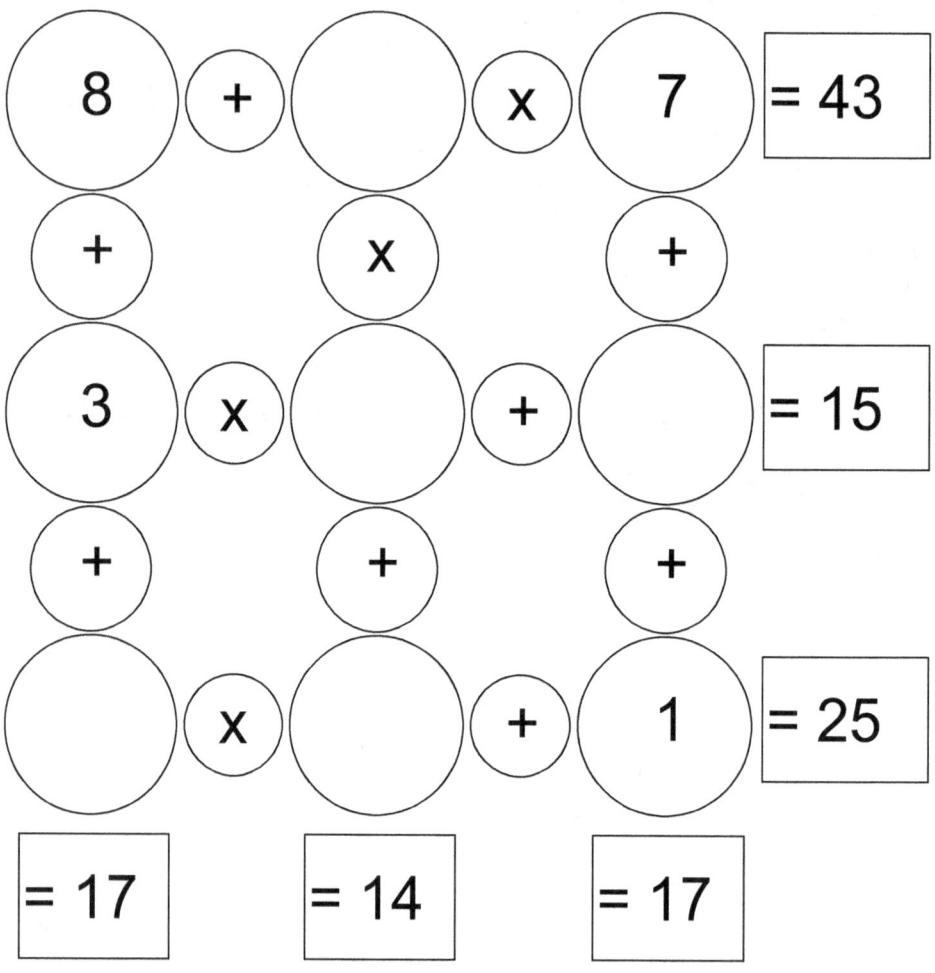

Solution at Page No 126

MATH IS FUN 5A

Puzzle Number 65

Digits to be used: 4,5,6,8,9

Note: Each Digit can be used only once

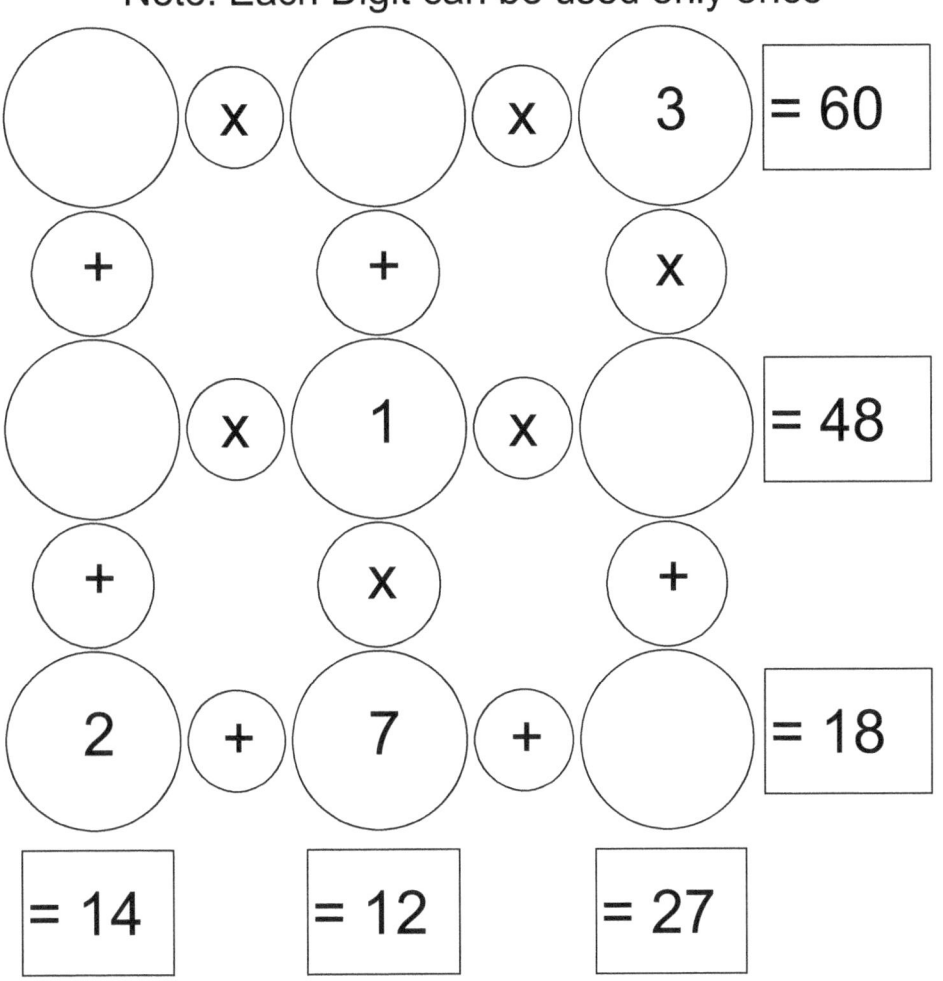

Solution at Page No 127

MATH IS FUN 5A

Puzzle Number 66

Digits to be used: 3,4,6,8,9

Note: Each Digit can be used only once

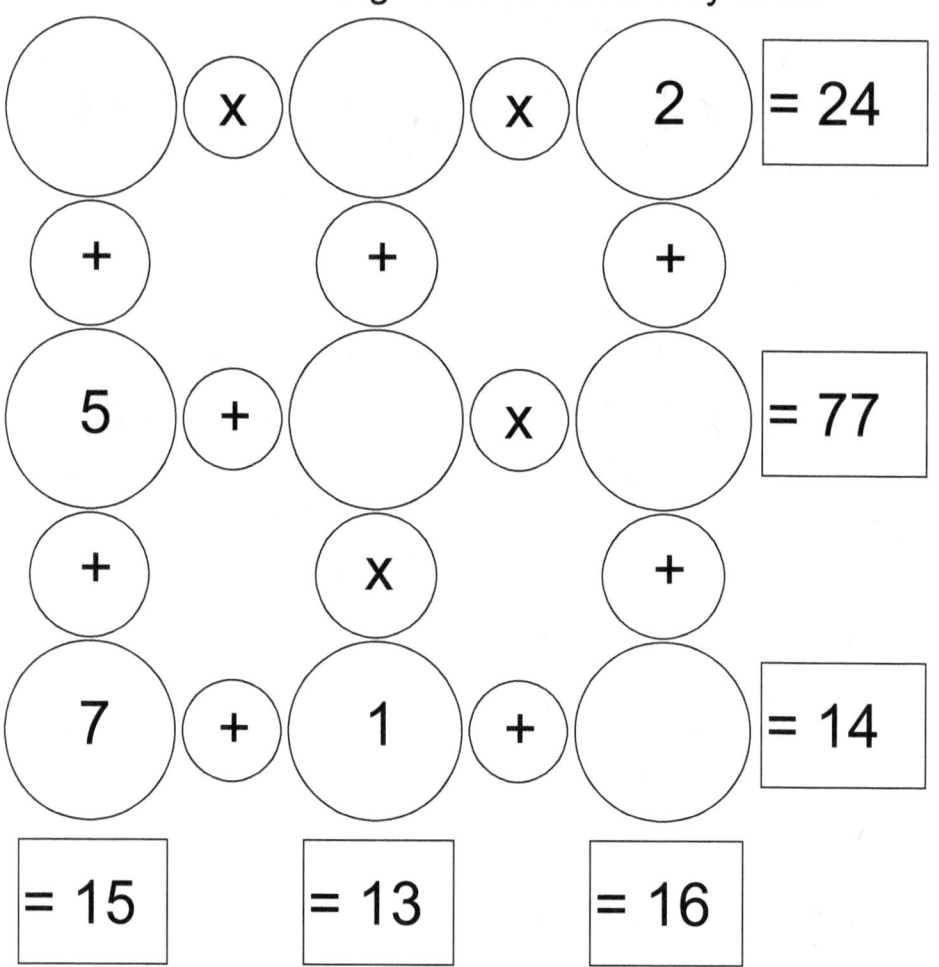

Solution at Page No 127

MATH IS FUN 5A

Puzzle Number 67

Digits to be used: 2,4,6,8,9

Note: Each Digit can be used only once

( ) + ( ) x (1) = 10

+   +   x

(3) x ( ) x ( ) = 108

x   x   +

( ) + (7) + (5) = 18

= 26   = 65   = 9

Solution at Page No 127

[77]

Puzzle Number 68

Digits to be used: 1,2,4,6,8

Note: Each Digit can be used only once

( ) x 5 + ( ) = 26

x + x

( ) + ( ) x 9 = 26

+ x +

( ) x 3 + 7 = 10

= 33   = 11   = 61

Solution at Page No 127

[78]

MATH IS FUN 5A

Puzzle Number 69

Digits to be used: 2,4,5,6,9

Note: Each Digit can be used only once

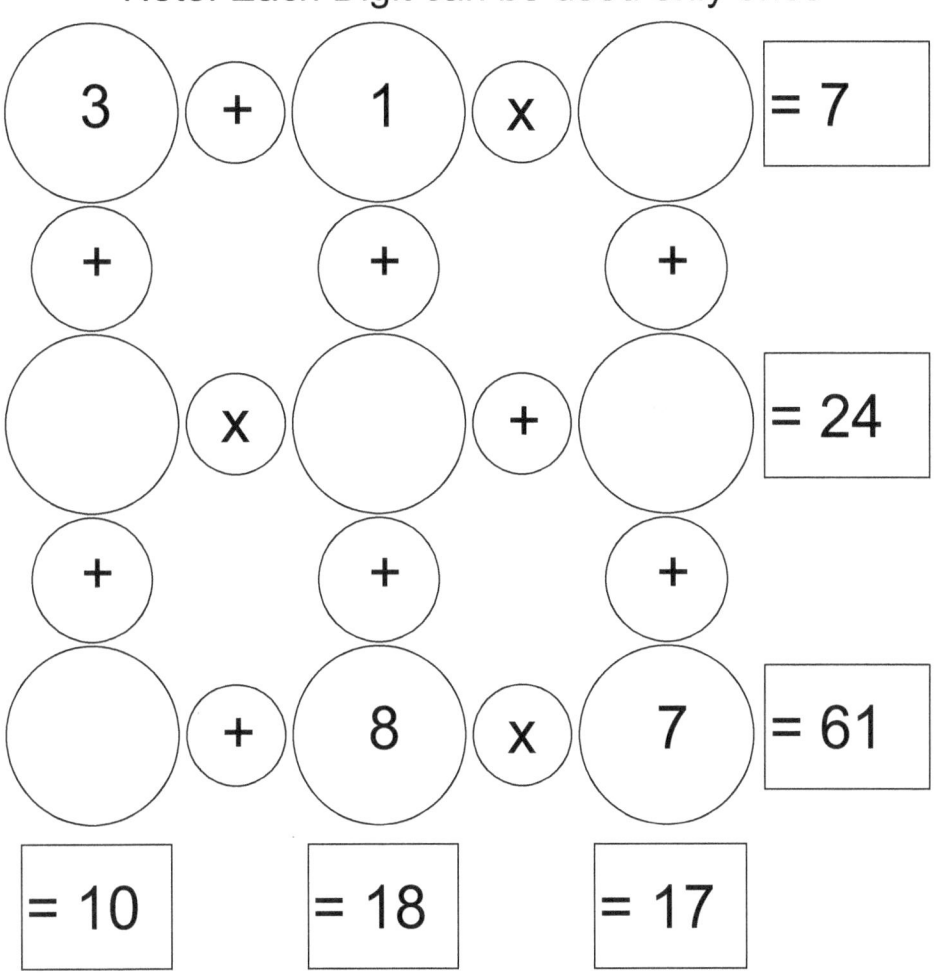

Solution at Page No 128

Puzzle Number 70

Digits to be used: 1,2,5,6,7

Note: Each Digit can be used only once

4 x ( ) + 3 = 7

x   x   +

( ) x ( ) + ( ) = 41

x   +   +

( ) x 9 + 8 = 26

= 56   = 14   = 17

Solution at Page No 128

Puzzle Number 71

Digits to be used: 1,3,5,6,8

Note: Each Digit can be used only once

( ) + (2) + (7) = 14

x   x   x

( ) + ( ) x (9) = 17

+   x   +

(4) x ( ) + ( ) = 27

= 44   = 12   = 66

Solution at Page No 128

MATH IS FUN 5A

Puzzle Number 72

Digits to be used: 1,2,3,4,9

Note: Each Digit can be used only once

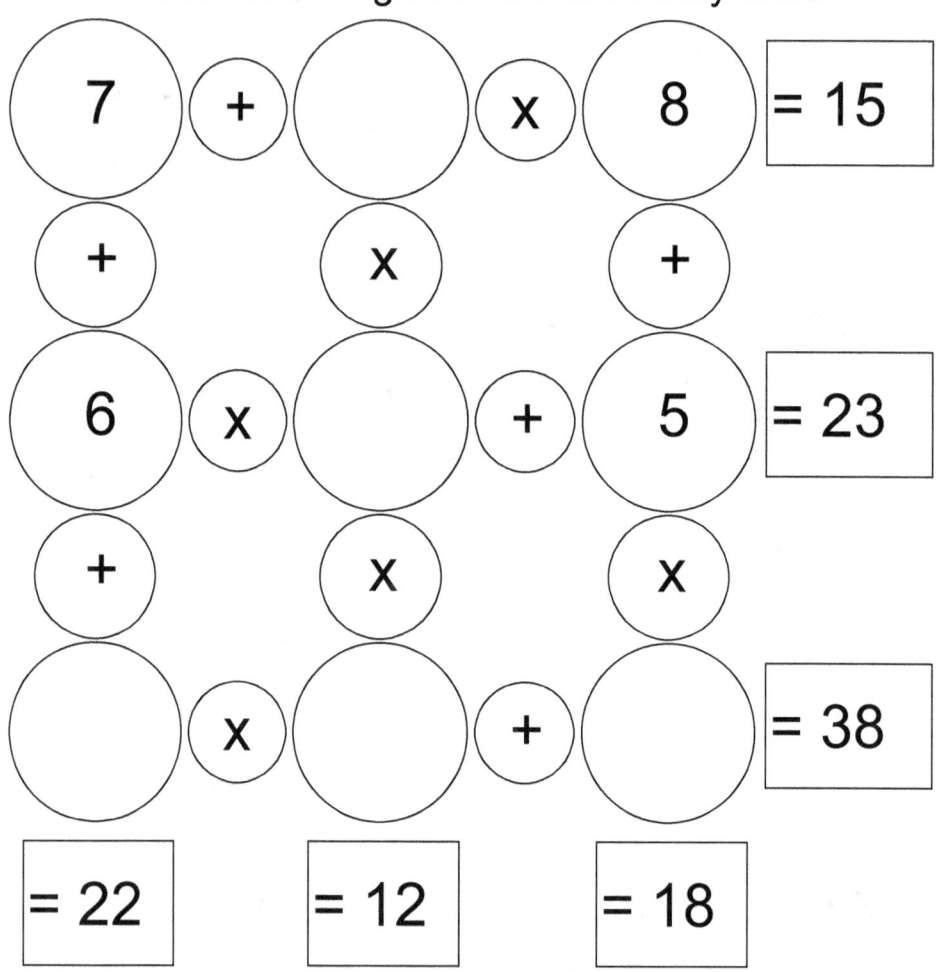

Solution at Page No 128

Puzzle Number 73

Digits to be used: 1,2,3,4,8

Note: Each Digit can be used only once

( ) X (9) X ( ) = 18

+   X   X

(5) X ( ) + ( ) = 44

+   +   +

(6) X ( ) + (7) = 25

= 12   = 75   = 15

Solution at Page No 129

[83]

Puzzle Number 74

Digits to be used: 3,5,7,8,9

Note: Each Digit can be used only once

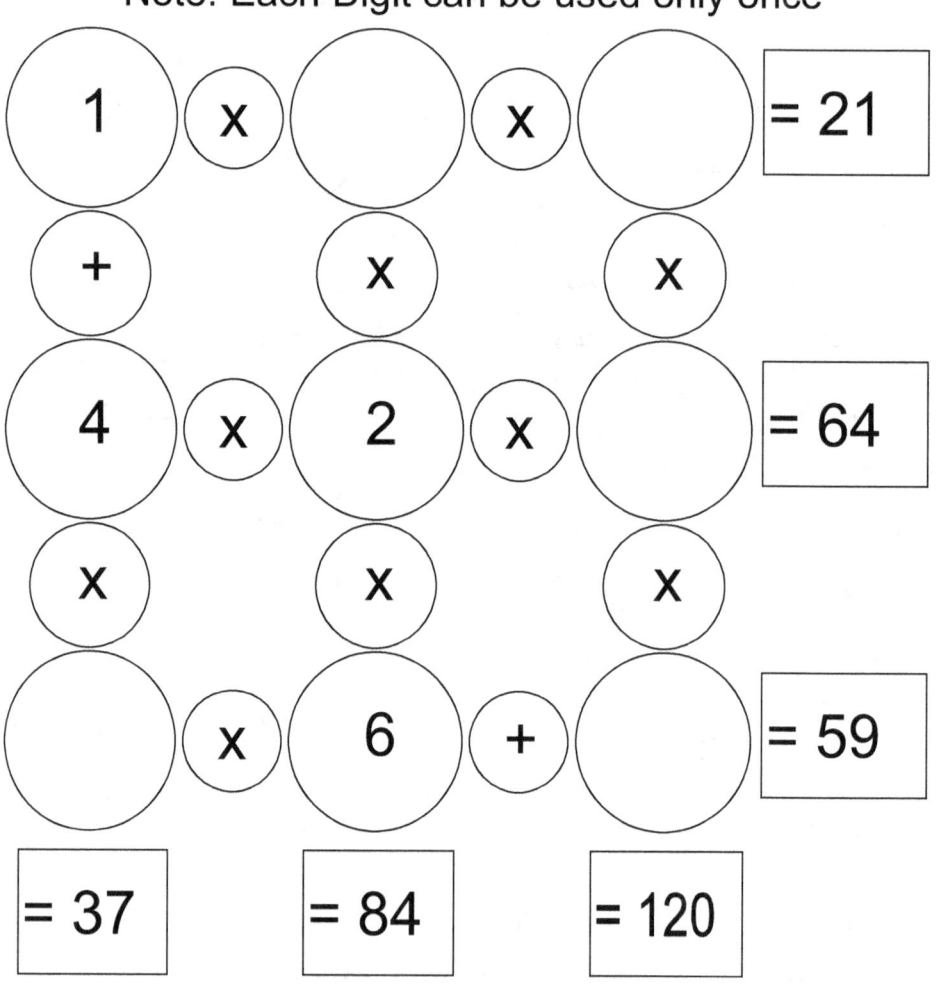

Solution at Page No 129

[84]

Puzzle Number 75

Digits to be used: 2,4,5,6,8

Note: Each Digit can be used only once

( ) + (9) x ( ) = 41

+   +   x

( ) x ( ) + (7) = 19

+   +   +

(1) x ( ) + (3) = 11

= 8   = 23   = 31

Solution at Page No 129

Puzzle Number 76

Digits to be used: 2,4,5,6,9

Note: Each Digit can be used only once

7	+	8	x ( ) = 47
+		+	+
1	x ( )	x ( )	= 36
x		x	x
( )	+	3	+ ( ) = 11
= 13		= 35	= 13

Solution at Page No 129

Puzzle Number 77

Digits to be used: 1,2,6,7,9

Note: Each Digit can be used only once

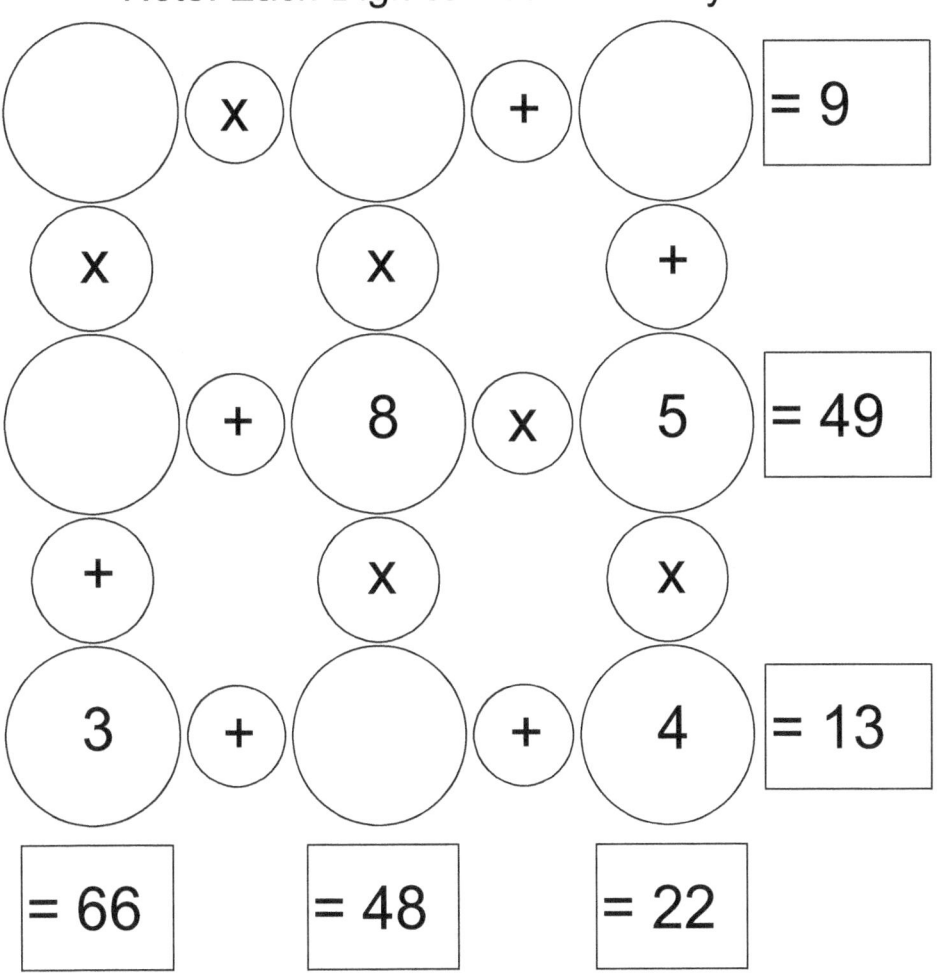

Solution at Page No 130

[87]

Puzzle Number 78

Digits to be used: 1,2,3,4,7

Note: Each Digit can be used only once

$\bigcirc$ + $\bigcirc$ + $\bigcirc$ = 11

+   x   +

6 + 5 x $\bigcirc$ = 26

x   +   x

9 x 8 + $\bigcirc$ = 74

= 55   = 43   = 11

Solution at Page No 130

[88]

Puzzle Number 79

Digits to be used: 1,2,3,4,7

Note: Each Digit can be used only once

( ) X	( ) +	( ) = 25
X	+	+
5 X	6 X	( ) = 60
X	+	X
( ) X	9 X	8 = 72
= 15	= 22	= 20

Solution at Page No 130

MATH IS FUN 5A

Puzzle Number 80

Digits to be used: 2,3,4,7,9

Note: Each Digit can be used only once

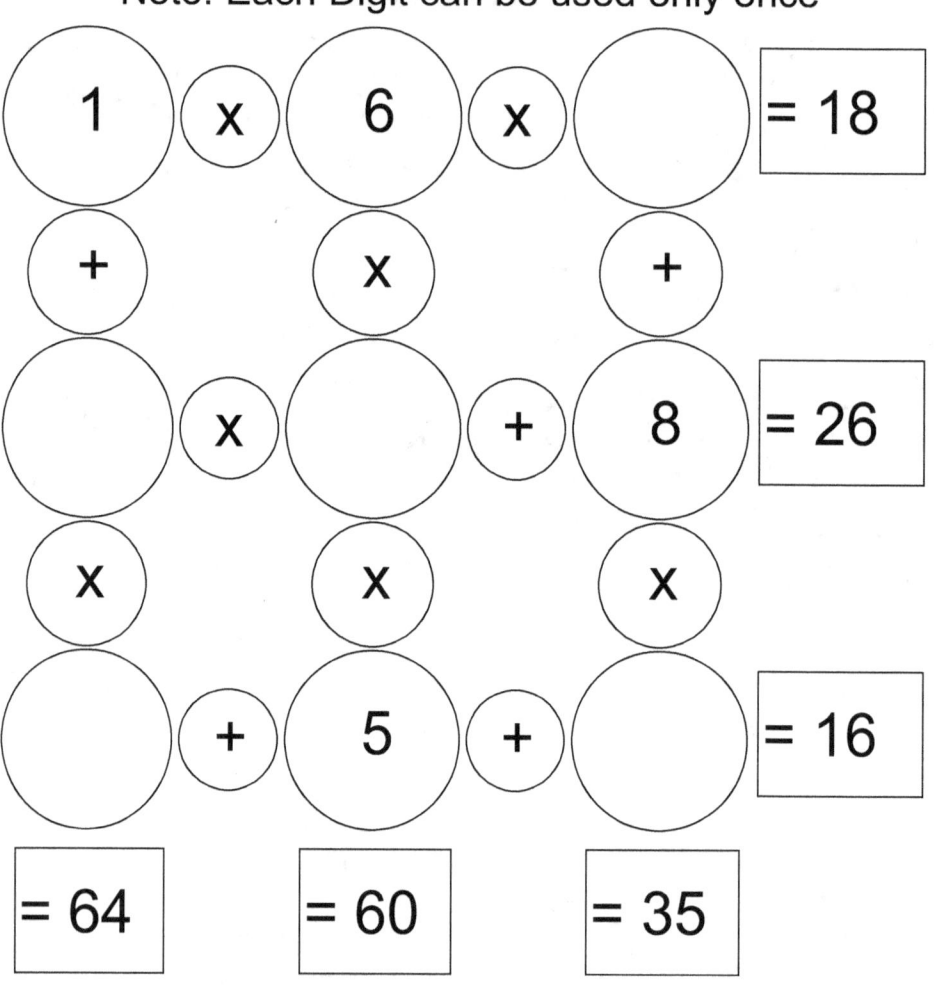

Solution at Page No 130

Puzzle Number 81

Digits to be used: 1,2,4,8,9

Note: Each Digit can be used only once

( ) + (3) x (5) = 23

+   +   +

(6) x ( ) + (7) = 31

+   +   x

( ) x ( ) x ( ) = 18

= 16   = 8   = 68

Solution at Page No 131

[91]

Puzzle Number 82

Digits to be used: 2,4,5,7,9

Note: Each Digit can be used only once

( ) x ( ) x (6) = 120

x   x   x

( ) x (1) + ( ) = 11

+   +   +

(3) x (8) + ( ) = 31

= 39   = 13   = 19

Solution at Page No 131

MATH IS FUN 5A

Puzzle Number 83

Digits to be used: 1,6,7,8,9

Note: Each Digit can be used only once

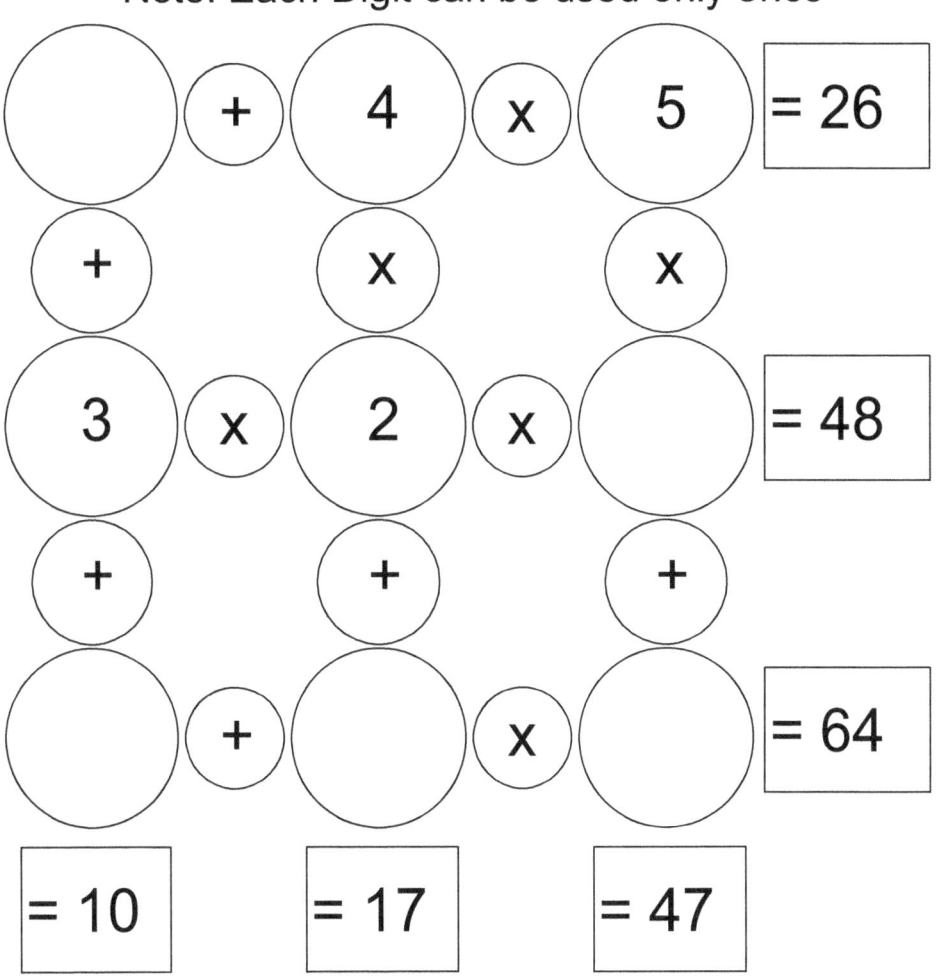

Solution at Page No 131

[93]

Puzzle Number 84

Digits to be used: 1,3,5,8,9

Note: Each Digit can be used only once

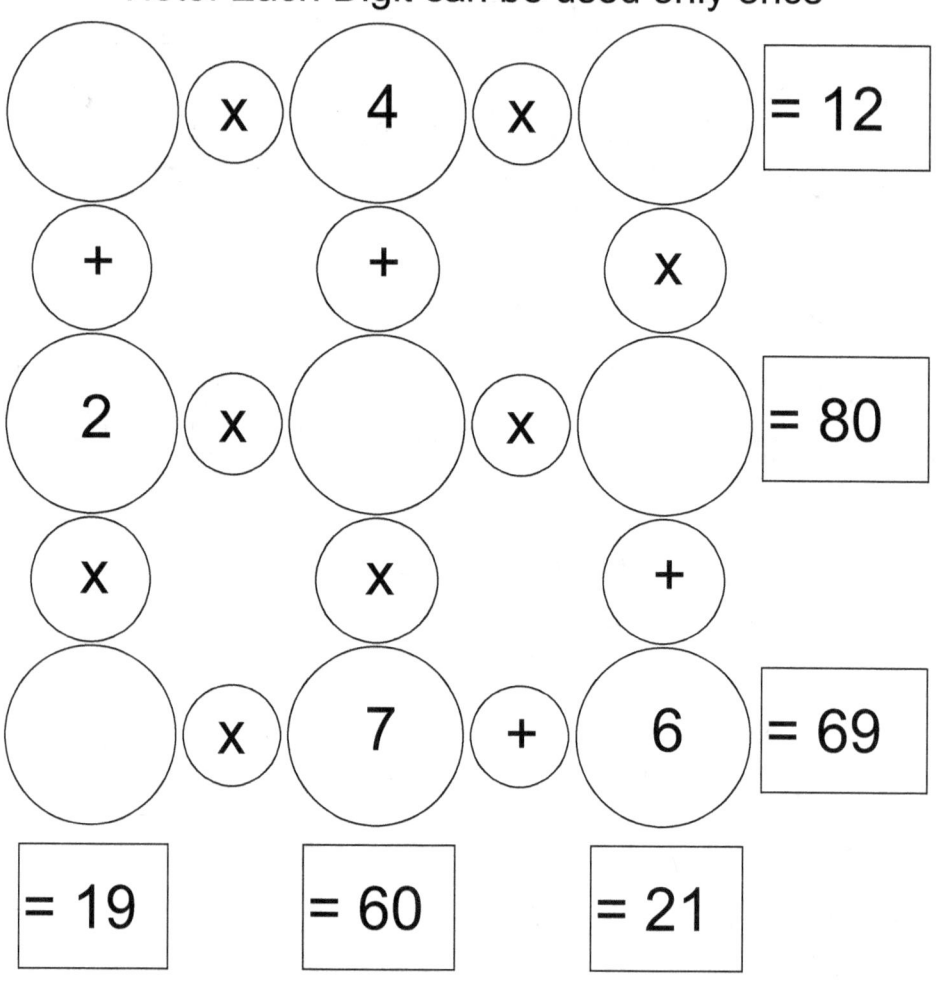

Solution at Page No 131

[94]

Puzzle Number 85

Digits to be used: 1,2,6,7,9

Note: Each Digit can be used only once

( ) x (8) + (4) = 20

+   x   +

(3) + ( ) + ( ) = 16

+   +   x

( ) x ( ) + (5) = 14

= 14   = 49   = 39

Solution at Page No 132

[95]

Puzzle Number 86

Digits to be used: 2,5,7,8,9

Note: Each Digit can be used only once

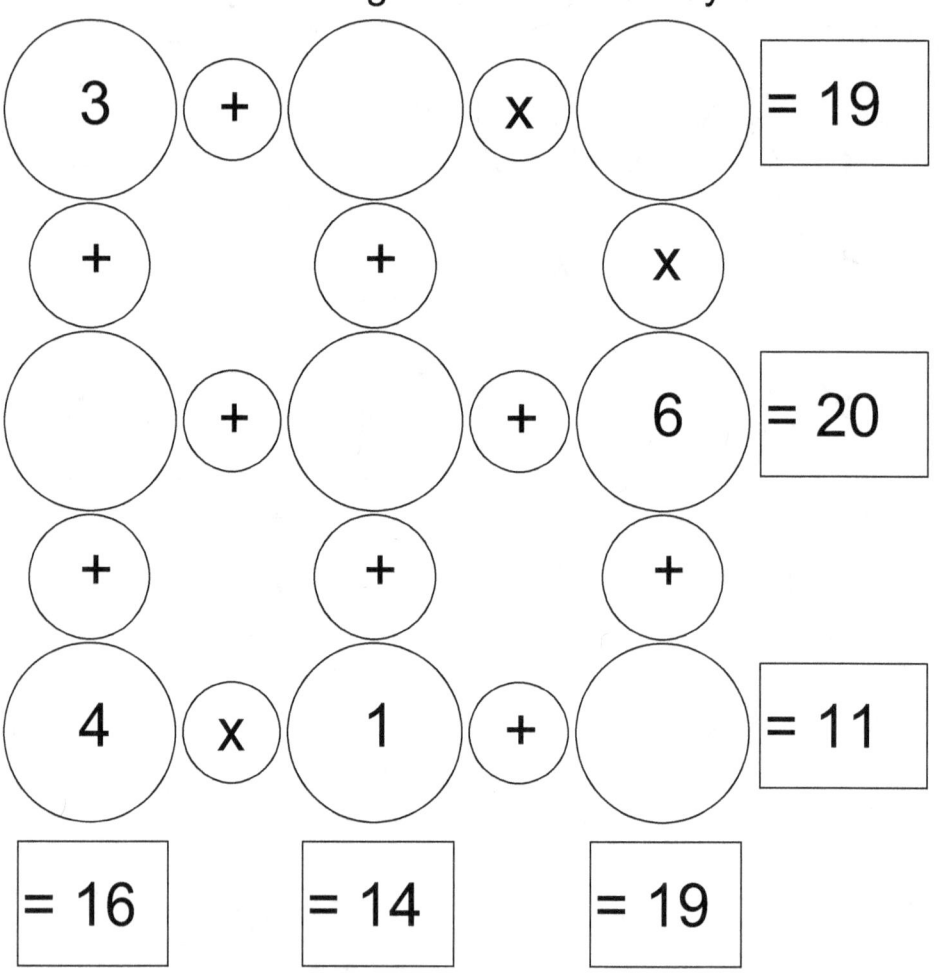

Solution at Page No 132

[96]

Puzzle Number 87

Digits to be used: 1,5,6,8,9

Note: Each Digit can be used only once

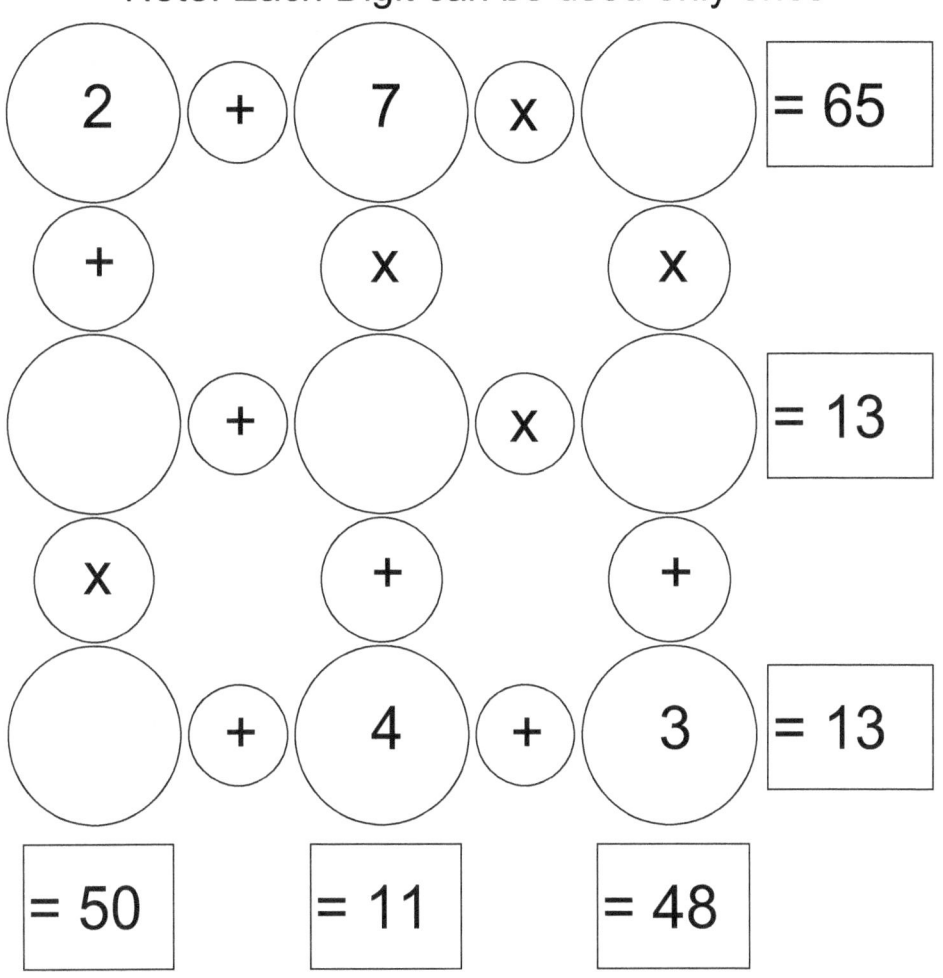

Solution at Page No 132

[97]

MATH IS FUN 5A

Puzzle Number 88

Digits to be used: 1,5,7,8,9

Note: Each Digit can be used only once

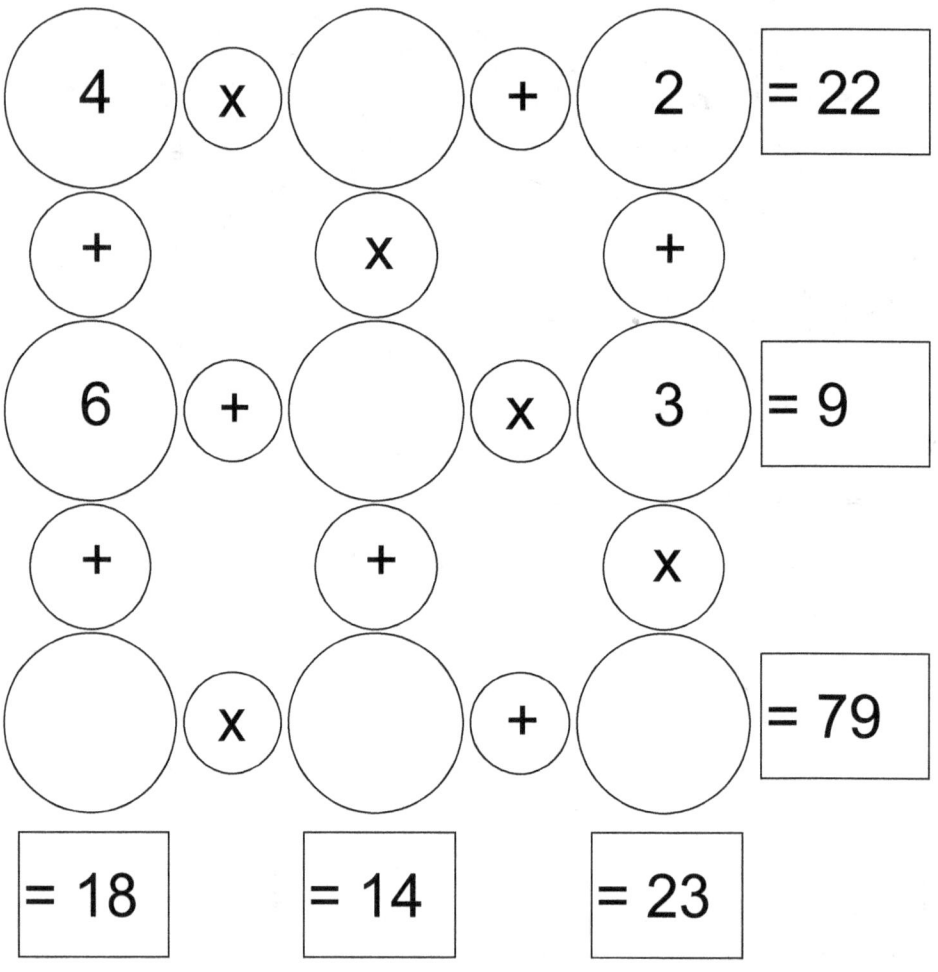

Solution at Page No 132

MATH IS FUN 5A

Puzzle Number 89

Digits to be used: 3,4,5,8,9

Note: Each Digit can be used only once

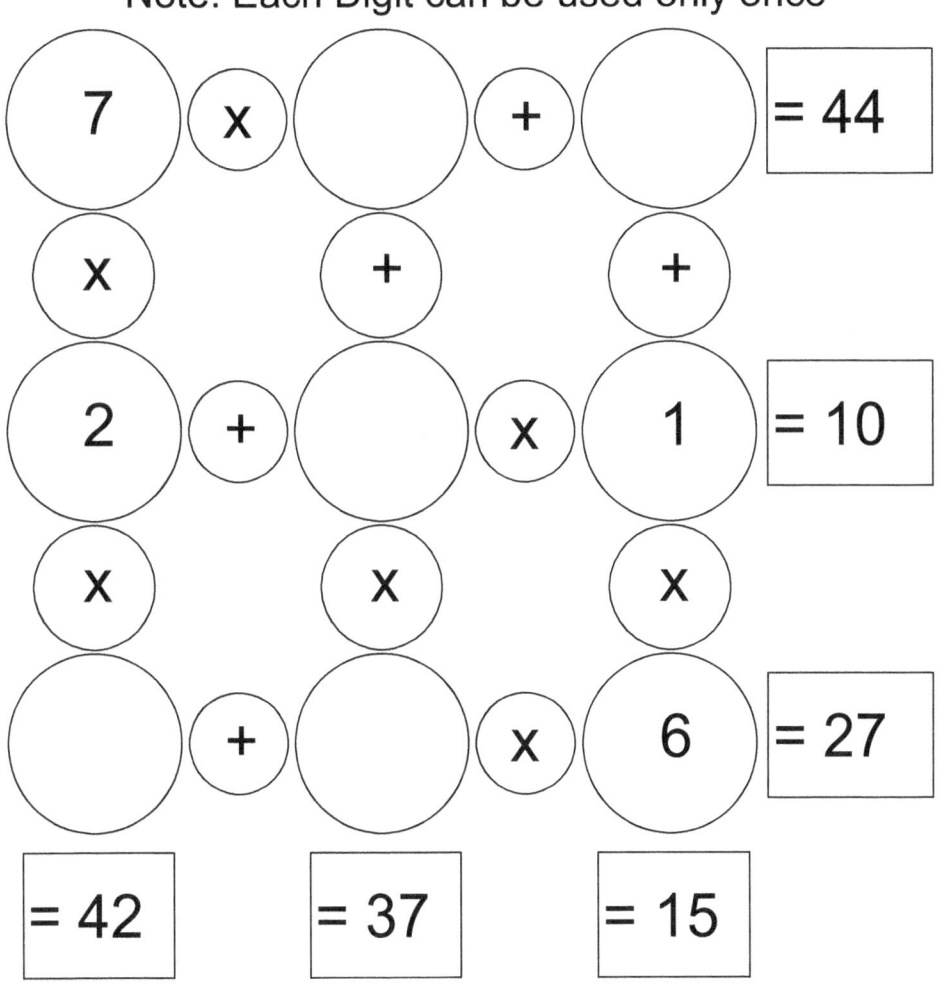

Solution at Page No 133

Puzzle Number 90

Digits to be used: 1,2,6,7,9

Note: Each Digit can be used only once

| 8 | + | 3 | + | | = 18 |

| + | | + | | x |

| | x | | x | | = 54 |

| x | | x | | + |

| 5 | + | 4 | + | | = 11 |

= 53   = 7   = 44

Solution at Page No 133

MATH IS FUN 5A

Puzzle Number 91

Digits to be used: 5,6,7,8,9

Note: Each Digit can be used only once

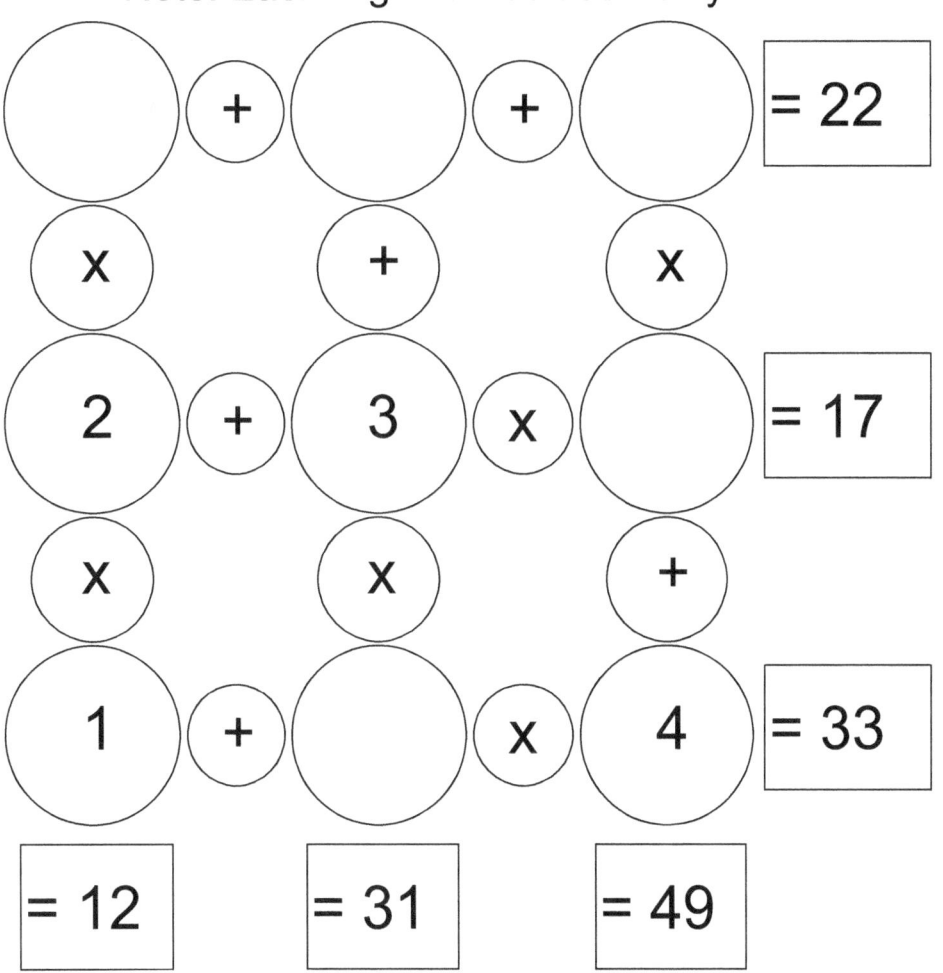

Solution at Page No 133

MATH IS FUN 5A

Puzzle Number 92

Digits to be used: 1,3,5,7,8

Note: Each Digit can be used only once

$2 + \bigcirc \times 6 = 20$

$+ \qquad + \qquad \times$

$\bigcirc + \bigcirc \times 4 = 39$

$\times \qquad + \qquad \times$

$9 + \bigcirc \times \bigcirc = 14$

$= 65 \qquad = 16 \qquad = 24$

Solution at Page No 133

MATH IS FUN 5A

Puzzle Number 93

Digits to be used: 1,4,5,7,9

Note: Each Digit can be used only once

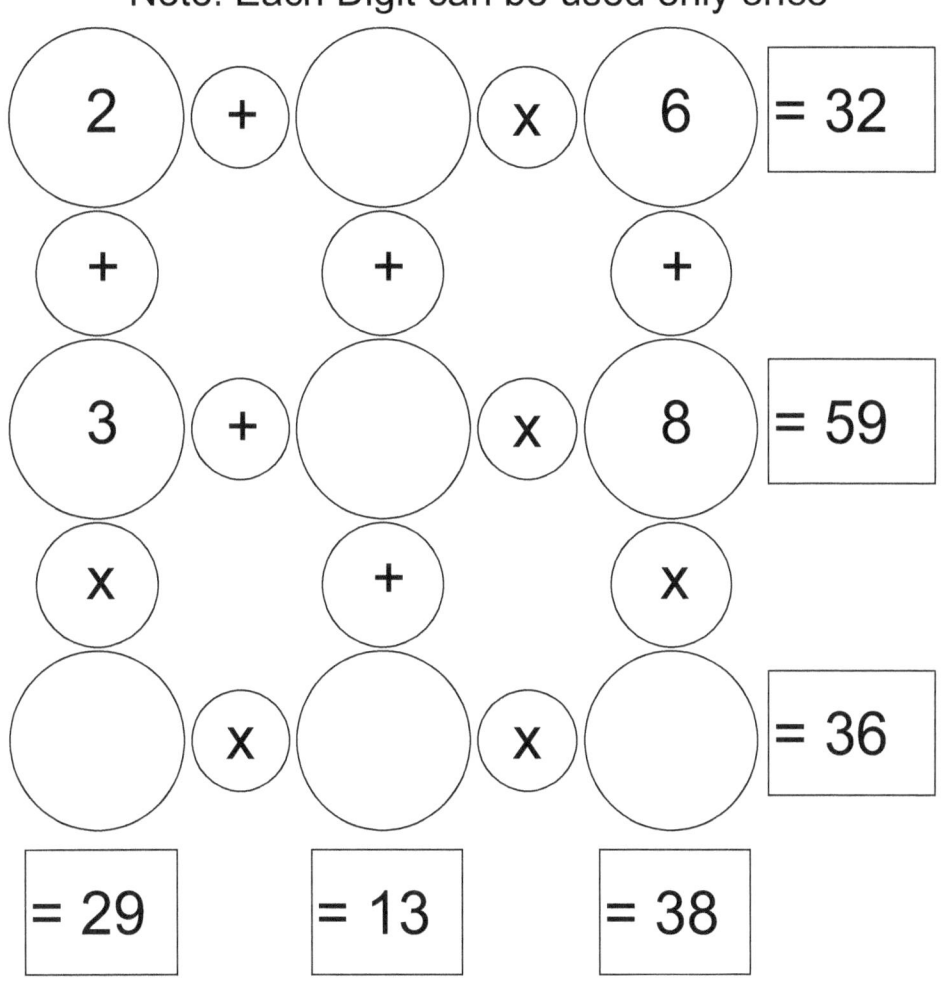

Solution at Page No 134

Puzzle Number 94

Digits to be used: 1,2,4,5,8

Note: Each Digit can be used only once

6 + ( ) + 7 = 18

+   +   +

( ) x ( ) + ( ) = 34

x   +   +

3 + ( ) x 9 = 12

= 30   = 10   = 18

Solution at Page No 134

[104]

MATH IS FUN 5A

Puzzle Number 95

Digits to be used: 1,2,5,7,8

Note: Each Digit can be used only once

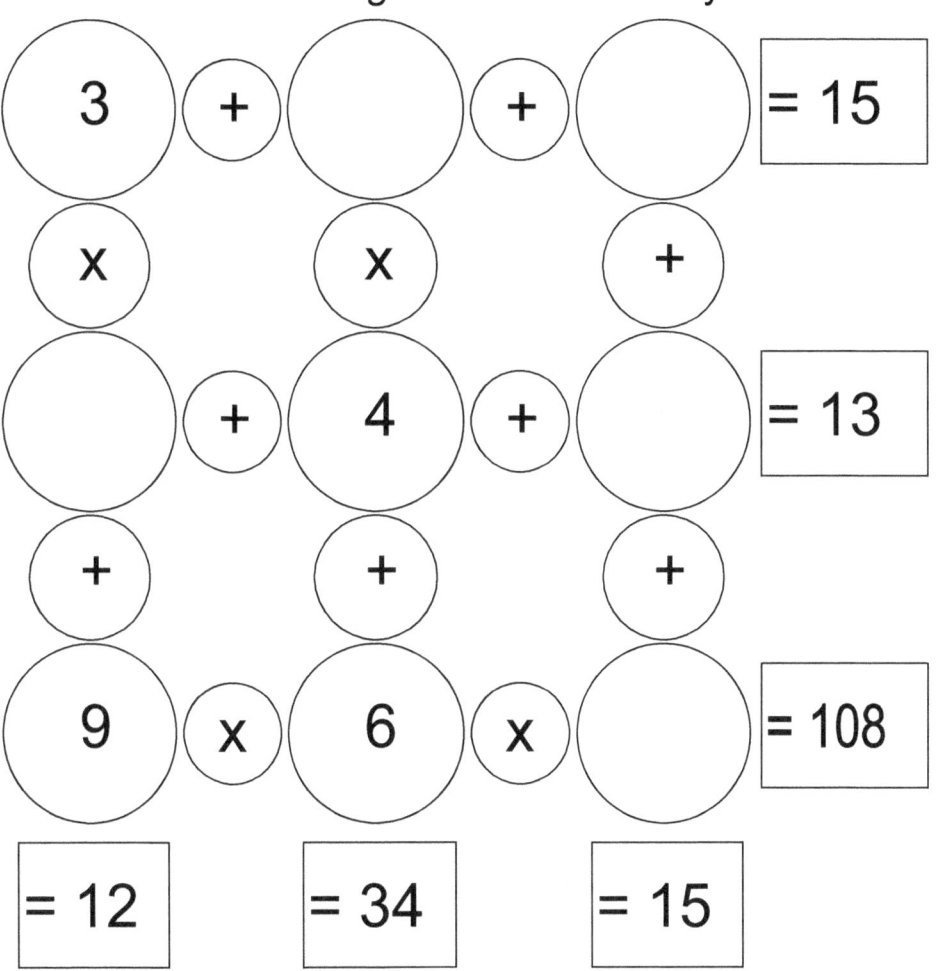

Solution at Page No 134

Puzzle Number 96

Digits to be used: 3,5,7,8,9

Note: Each Digit can be used only once

Solution at Page No 134

Puzzle Number 97

Digits to be used: 4,5,6,7,9

Note: Each Digit can be used only once

( )	+	(2)	+	(1)	= 7
+		+		+	
( )	+	( )	+	(3)	= 16
×		+		×	
(8)	+	( )	×	( )	= 53
= 52		= 14		= 28	

Solution at Page No 135

[107]

MATH IS FUN 5A

Puzzle Number 98

Digits to be used: 1,4,6,7,8

Note: Each Digit can be used only once

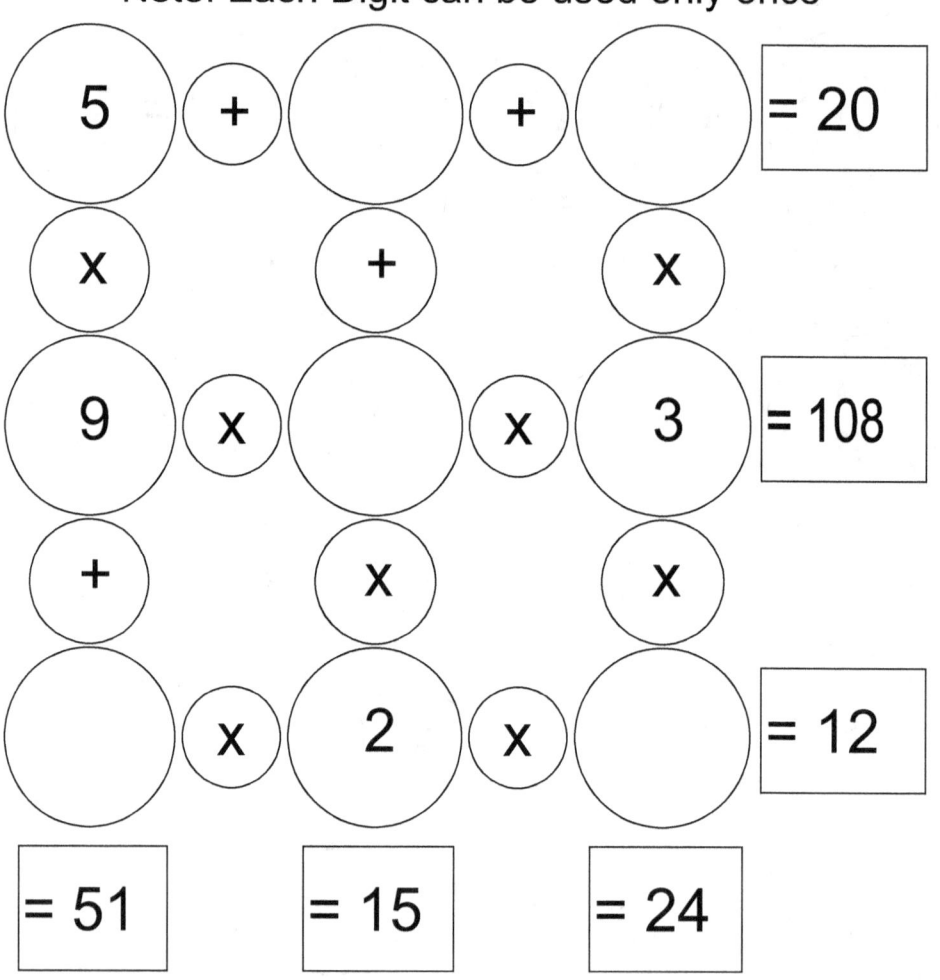

Solution at Page No 135

[108]

Puzzle Number 99

Digits to be used: 1,2,5,6,9

Note: Each Digit can be used only once

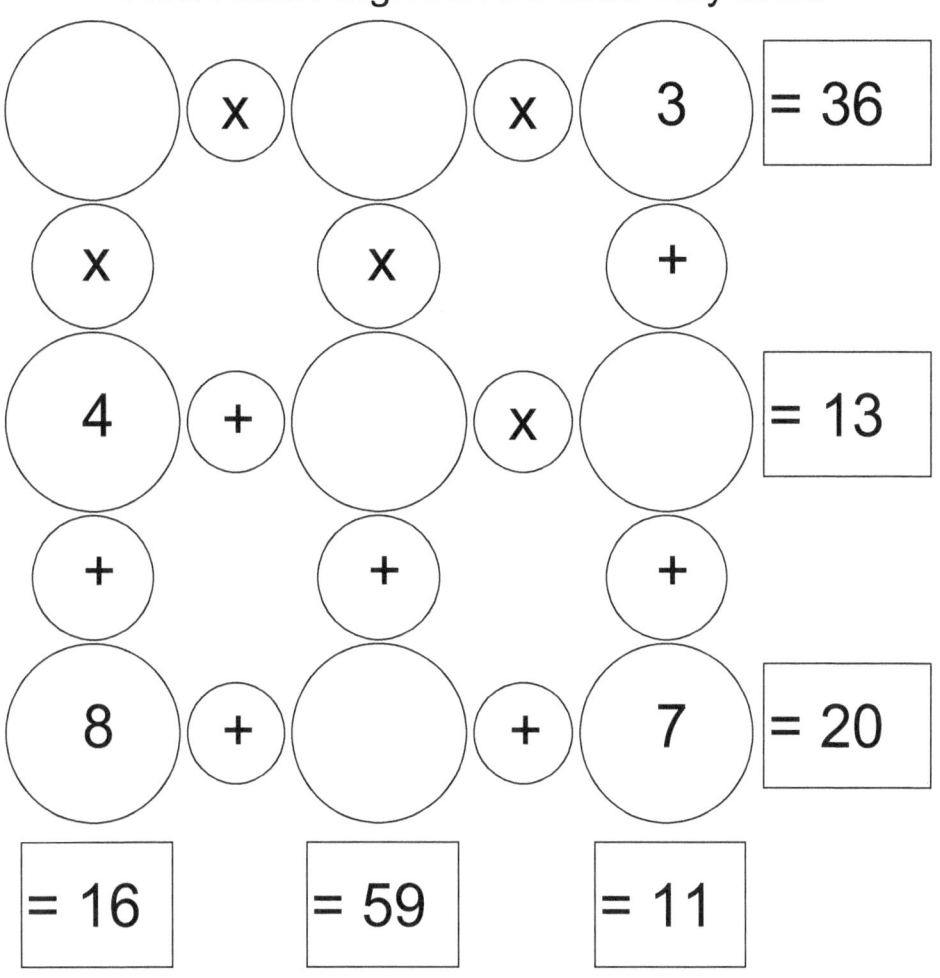

Solution at Page No 135

MATH IS FUN 5A

Puzzle Number 100

Digits to be used: 1,3,4,5,6

Note: Each Digit can be used only once

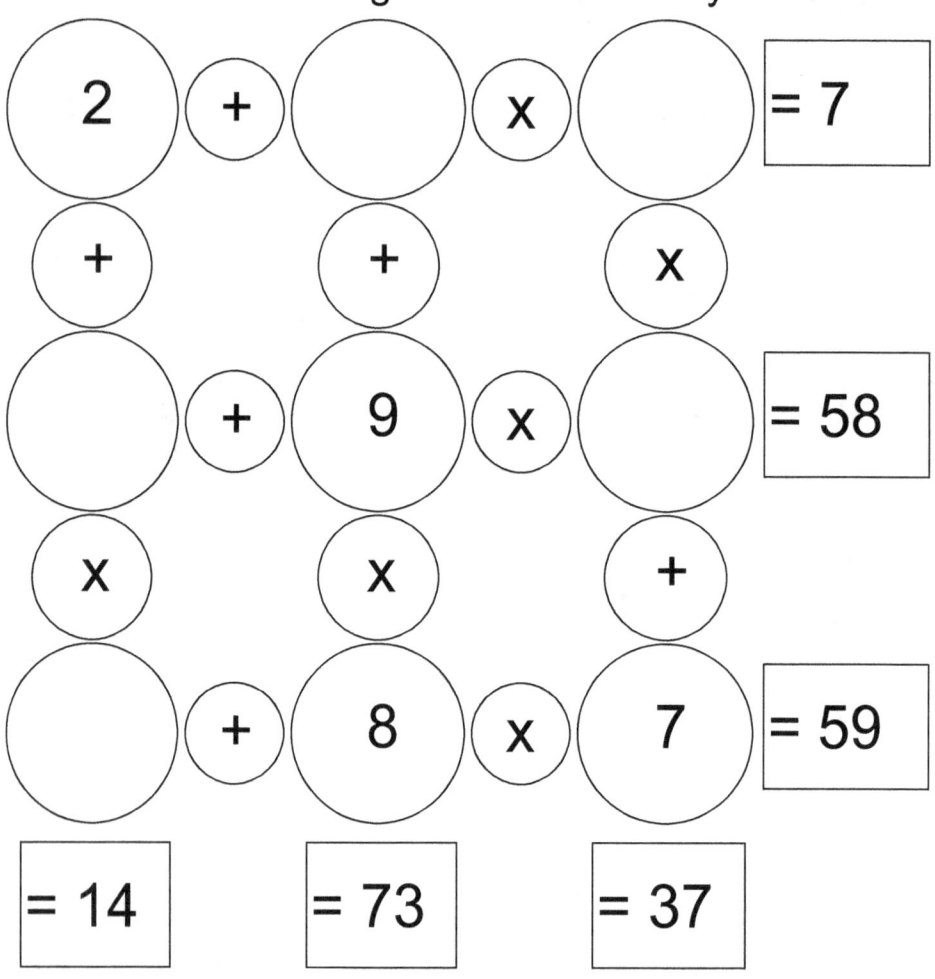

Solution at Page No 135

MATH IS FUN 5A

Solution of Puzzle 1

6 + 2 + 1 = 9
x   x   x
5 x 8 + 3 = 43
+   x   x
9 + 4 + 7 = 20
= 39 = 64 = 21

Solution of Puzzle 2

2 x 9 x 4 = 72
x   x   +
8 x 1 + 7 = 15
+   +   x
5 + 6 + 3 = 14
= 21 = 15 = 25

Solution of Puzzle 3

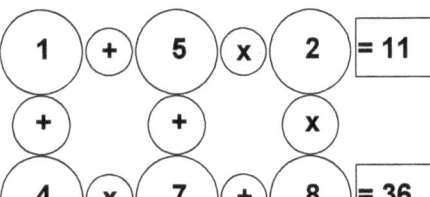

1 + 5 x 2 = 11
+   +   x
4 x 7 + 8 = 36
+   x   x
9 + 3 + 6 = 18
= 14 = 26 = 96

Solution of Puzzle 4

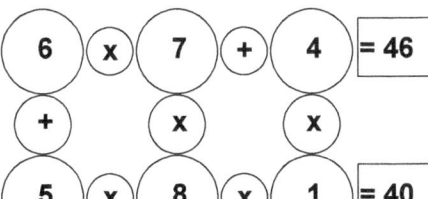

6 x 7 + 4 = 46
+   x   x
5 x 8 x 1 = 40
x   +   +
2 x 9 x 3 = 54
= 16 = 65 = 7

[111]

MATH IS FUN 5A

Solution of Puzzle 5

```
4  +  3  x  6  = 22
x     +     +
9  x  8  +  2  = 74
+     +     +
7  +  1  x  5  = 12
=43   =12    =13
```

Solution of Puzzle 6

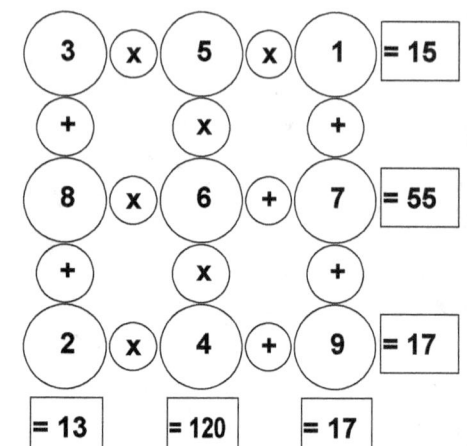

Solution of Puzzle 7

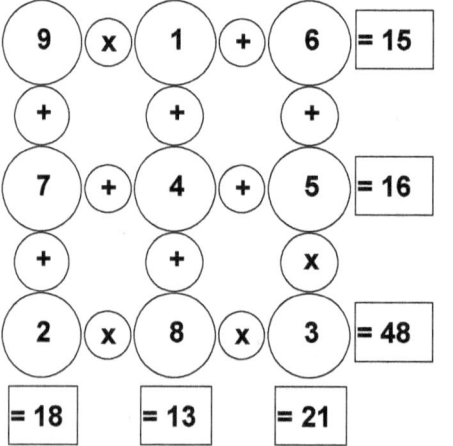

Solution of Puzzle 8

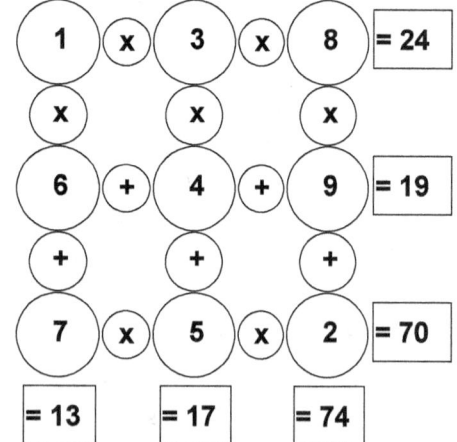

[112]

MATH IS FUN 5A

Solution of Puzzle 9

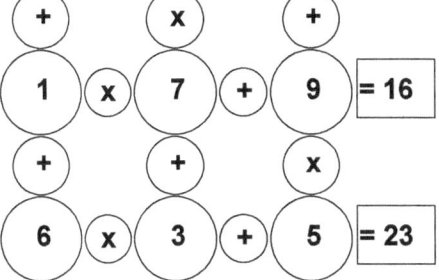

4 + 8 x 2	= 20
1 x 7 + 9	= 16
6 x 3 + 5	= 23
= 11	= 59	= 47

Solution of Puzzle 10

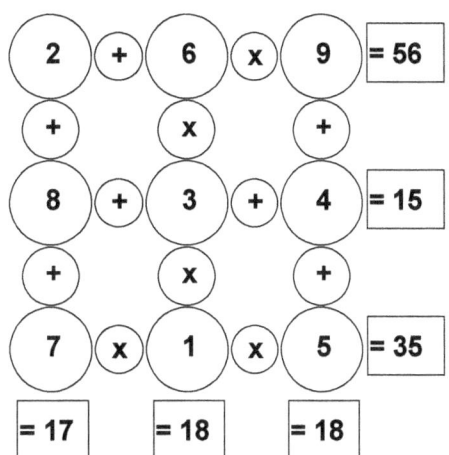

2 + 6 x 9	= 56
8 + 3 + 4	= 15
7 x 1 x 5	= 35
= 17	= 18	= 18

Solution of Puzzle 11

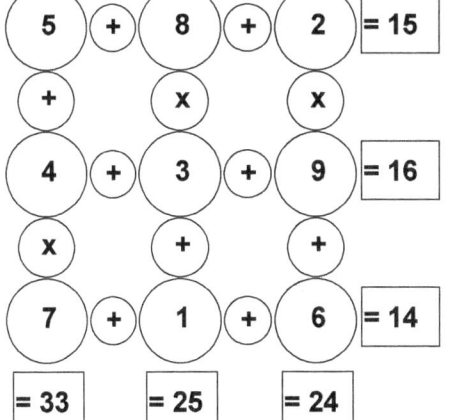

5 + 8 + 2	= 15
4 + 3 + 9	= 16
7 + 1 + 6	= 14
= 33	= 25	= 24

Solution of Puzzle 12

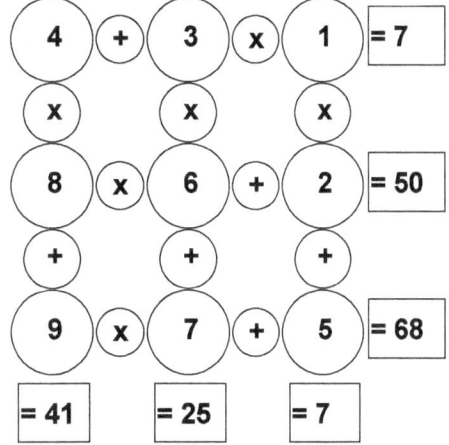

4 + 3 x 1	= 7
8 x 6 + 2	= 50
9 x 7 + 5	= 68
= 41	= 25	= 7

[113]

MATH IS FUN 5A

Solution of Puzzle 13

```
( 3 ) + ( 7 ) + ( 1 ) = 11
  +       +       +
( 9 ) x ( 5 ) + ( 2 ) = 47
  +       +       x
( 6 ) + ( 4 ) + ( 8 ) = 18
= 18    = 16    = 17
```

Solution of Puzzle 14

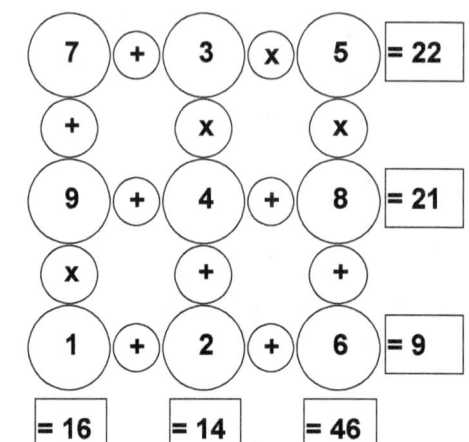

```
( 7 ) + ( 3 ) x ( 5 ) = 22
  +       x       x
( 9 ) + ( 4 ) + ( 8 ) = 21
  x       +       +
( 1 ) + ( 2 ) + ( 6 ) = 9
= 16    = 14    = 46
```

Solution of Puzzle 15

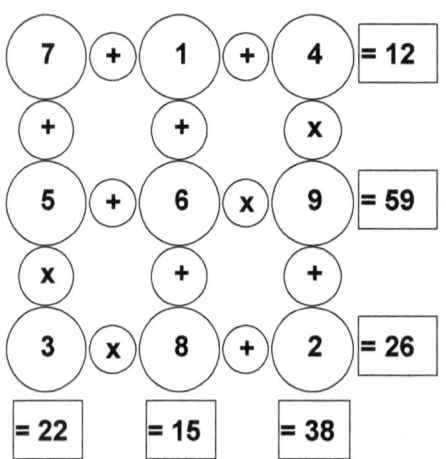

```
( 7 ) + ( 1 ) + ( 4 ) = 12
  +       +       x
( 5 ) + ( 6 ) x ( 9 ) = 59
  x       +       +
( 3 ) x ( 8 ) + ( 2 ) = 26
= 22    = 15    = 38
```

Solution of Puzzle 16

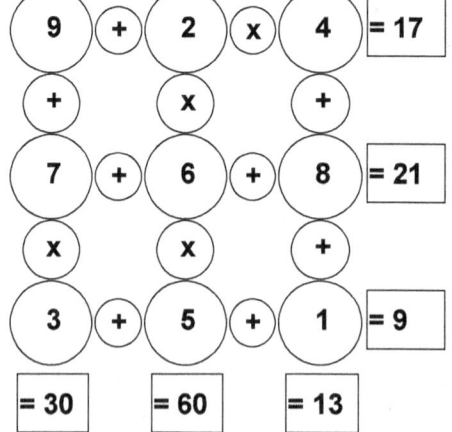

```
( 9 ) + ( 2 ) x ( 4 ) = 17
  +       x       +
( 7 ) + ( 6 ) + ( 8 ) = 21
  x       x       +
( 3 ) + ( 5 ) + ( 1 ) = 9
= 30    = 60    = 13
```

[114]

MATH IS FUN 5A

Solution of Puzzle 17

7	+	3	x	8	= 31
x		x		x
2	x	6	+	9	= 21
+		x		+
1	x	5	+	4	= 9
= 15		= 90		= 76

Solution of Puzzle 18

2	x	7	+	9	= 23
x		+		+
5	x	3	+	6	= 21
x		+		x
8	+	4	x	1	= 12
= 80		= 14		= 15

Solution of Puzzle 19

7	x	2	+	4	= 18
x		+		+
1	x	9	+	5	= 14
+		+		x
8	+	3	+	6	= 17
= 15		= 14		= 34

Solution of Puzzle 20

4	x	6	+	7	= 31
+		+		+
8	+	9	+	3	= 20
x		+		x
1	x	2	x	5	= 10
= 12		= 17		= 22

[115]

MATH IS FUN 5A

Solution of Puzzle 21

```
( 1 ) x ( 4 ) + ( 6 ) = 10
  x       +       +
( 3 ) x ( 9 ) + ( 8 ) = 35
  x       +       +
( 5 ) + ( 7 ) + ( 2 ) = 14
= 15    = 20    = 16
```

Solution of Puzzle 22

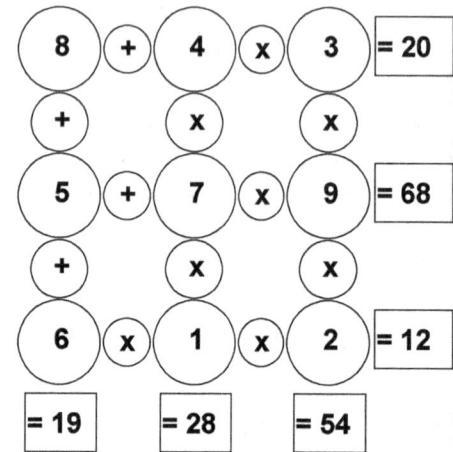

```
( 8 ) + ( 4 ) x ( 3 ) = 20
  +       x       x
( 5 ) + ( 7 ) x ( 9 ) = 68
  +       x       x
( 6 ) x ( 1 ) x ( 2 ) = 12
= 19    = 28    = 54
```

Solution of Puzzle 23

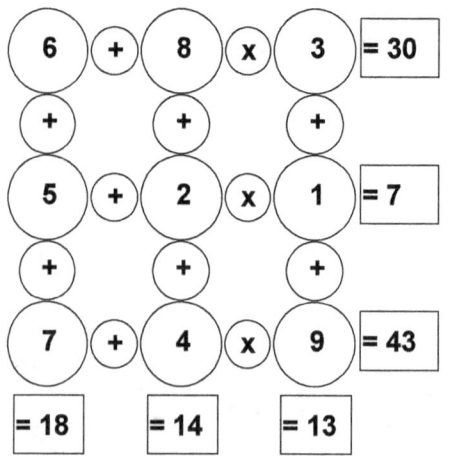

```
( 6 ) + ( 8 ) x ( 3 ) = 30
  +       +       +
( 5 ) + ( 2 ) x ( 1 ) = 7
  +       +       +
( 7 ) + ( 4 ) x ( 9 ) = 43
= 18    = 14    = 13
```

Solution of Puzzle 24

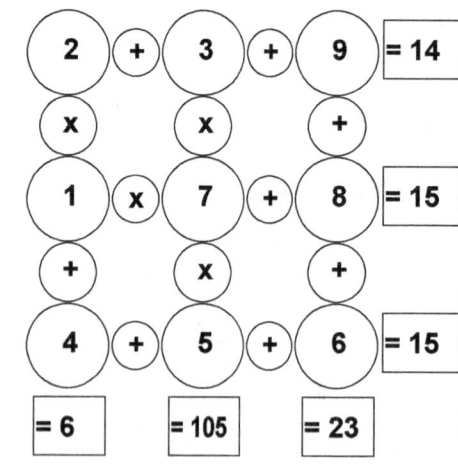

```
( 2 ) + ( 3 ) + ( 9 ) = 14
  x       x       +
( 1 ) x ( 7 ) + ( 8 ) = 15
  +       x       +
( 4 ) + ( 5 ) + ( 6 ) = 15
= 6     = 105   = 23
```

[116]

MATH IS FUN 5A

Solution of Puzzle 25

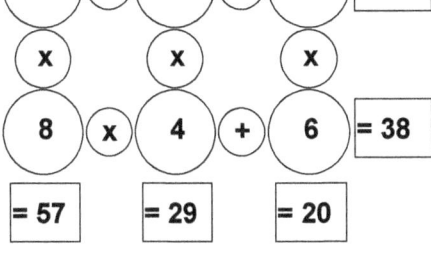

1	x	9	+	2	= 11
+		+		+	
7	x	5	x	3	= 105
x		x		x	
8	x	4	+	6	= 38
= 57		= 29		= 20	

Solution of Puzzle 26

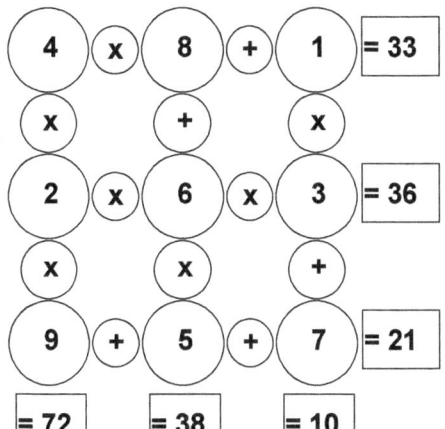

4	x	8	+	1	= 33
x		+		x	
2	x	6	x	3	= 36
x		x		+	
9	+	5	+	7	= 21
= 72		= 38		= 10	

Solution of Puzzle 27

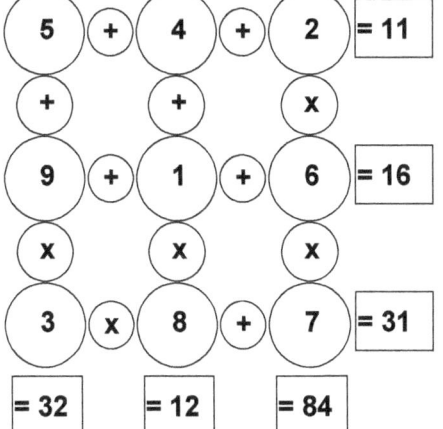

5	+	4	+	2	= 11
+		+		x	
9	+	1	+	6	= 16
x		x		x	
3	x	8	+	7	= 31
= 32		= 12		= 84	

Solution of Puzzle 28

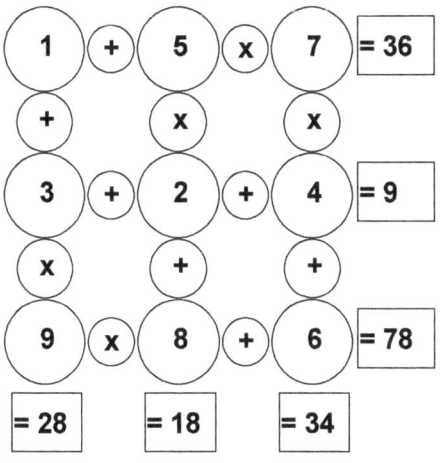

1	+	5	x	7	= 36
+		x		x	
3	+	2	+	4	= 9
x		+		+	
9	x	8	+	6	= 78
= 28		= 18		= 34	

[117]

MATH IS FUN 5A

Solution of Puzzle 29

```
( 3 ) + ( 9 ) + ( 5 ) = 17
  +       +       +
( 8 ) x ( 6 ) + ( 1 ) = 49
  +       +       +
( 4 ) + ( 2 ) + ( 7 ) = 13
 = 15    = 17    = 13
```

Solution of Puzzle 30

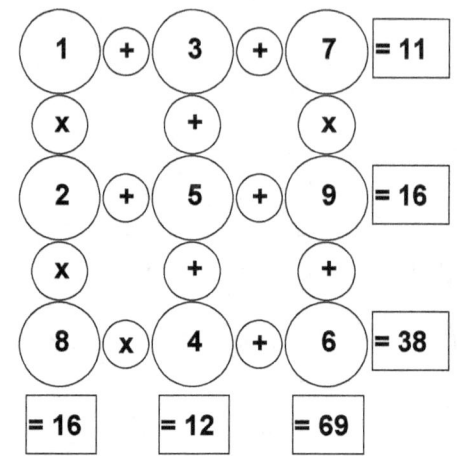

Solution of Puzzle 31

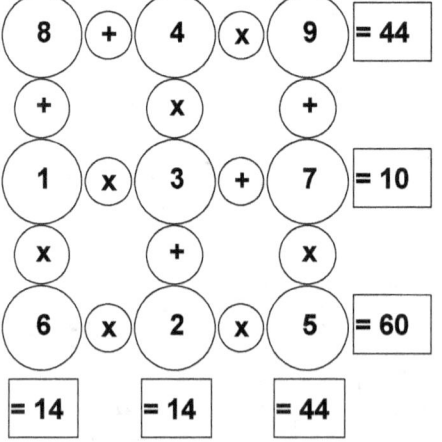

Solution of Puzzle 32

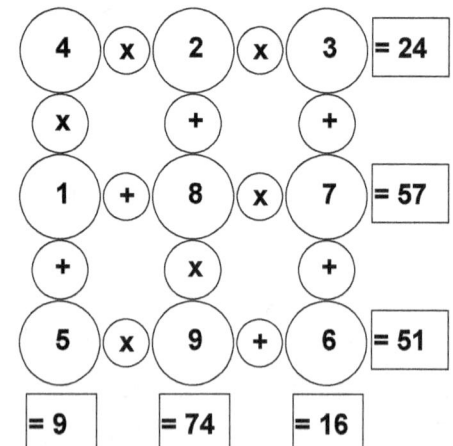

[118]

MATH IS FUN 5A

Solution of Puzzle 33

6 + 5 x 1	= 11
+ x +	
7 x 8 + 9	= 65
x + x	
4 + 2 x 3	= 10
= 34 = 42 = 28	

Solution of Puzzle 34

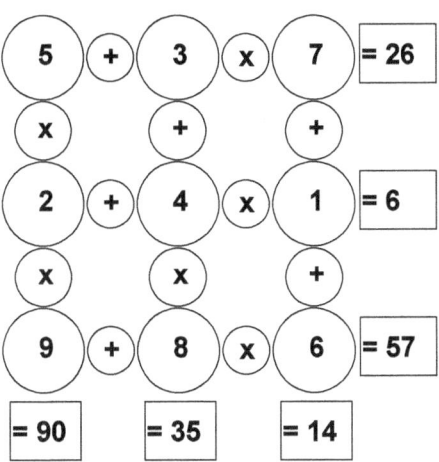

5 + 3 x 7	= 26
x + +	
2 + 4 x 1	= 6
x x +	
9 + 8 x 6	= 57
= 90 = 35 = 14	

Solution of Puzzle 35

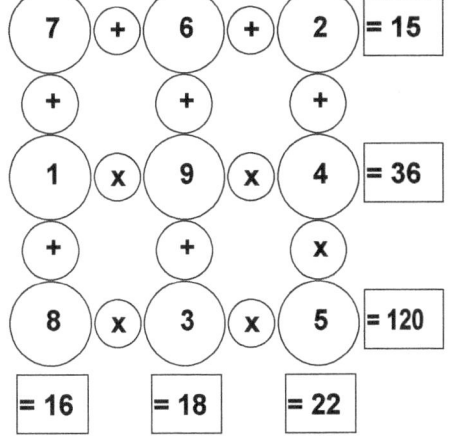

7 + 6 + 2	= 15
+ + +	
1 x 9 x 4	= 36
+ + x	
8 x 3 x 5	= 120
= 16 = 18 = 22	

Solution of Puzzle 36

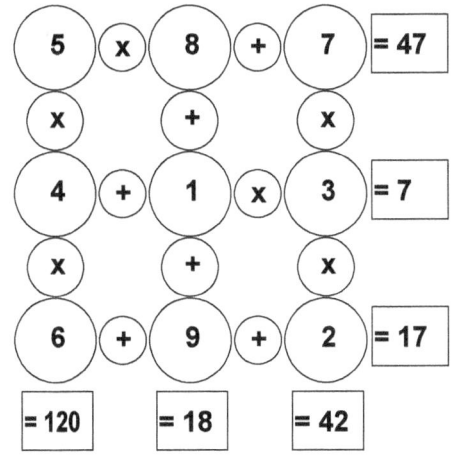

5 x 8 + 7	= 47
x + x	
4 + 1 x 3	= 7
x + x	
6 + 9 + 2	= 17
= 120 = 18 = 42	

[119]

MATH IS FUN 5A

Solution of Puzzle 37

```
( 2 ) x ( 4 ) x ( 7 ) = 56
  x       +       x
( 9 ) + ( 8 ) x ( 3 ) = 33
  x       +       +
( 1 ) x ( 6 ) + ( 5 ) = 11
= 18    = 18    = 26
```

Solution of Puzzle 38

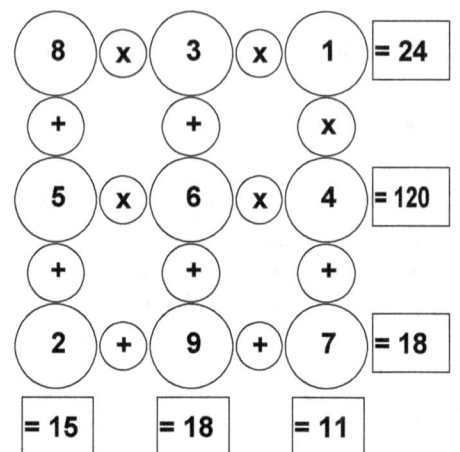

```
( 8 ) x ( 3 ) x ( 1 ) = 24
  +       +       x
( 5 ) x ( 6 ) x ( 4 ) = 120
  +       +       +
( 2 ) + ( 9 ) + ( 7 ) = 18
= 15    = 18    = 11
```

Solution of Puzzle 39

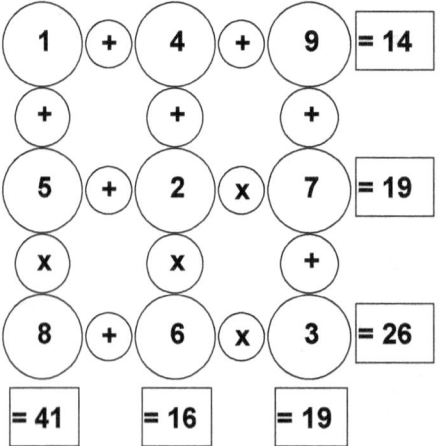

```
( 1 ) + ( 4 ) + ( 9 ) = 14
  +       +       +
( 5 ) + ( 2 ) x ( 7 ) = 19
  x       x       +
( 8 ) + ( 6 ) x ( 3 ) = 26
= 41    = 16    = 19
```

Solution of Puzzle 40

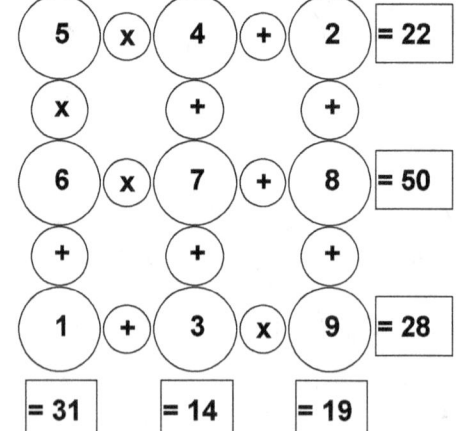

```
( 5 ) x ( 4 ) + ( 2 ) = 22
  x       +       +
( 6 ) x ( 7 ) + ( 8 ) = 50
  +       +       +
( 1 ) + ( 3 ) x ( 9 ) = 28
= 31    = 14    = 19
```

MATH IS FUN 5A

Solution of Puzzle 41

$$4 \times 3 + 9 = 21$$
$$\times \qquad + \qquad \times$$
$$1 + 6 + 2 = 9$$
$$\times \qquad \times \qquad \times$$
$$7 \times 8 + 5 = 61$$
$$= 28 \qquad = 51 \qquad = 90$$

Solution of Puzzle 42

$$4 + 6 + 9 = 19$$
$$\times \qquad \times \qquad +$$
$$5 + 7 + 8 = 20$$
$$\times \qquad + \qquad +$$
$$3 \times 1 + 2 = 5$$
$$= 60 \qquad = 43 \qquad = 19$$

Solution of Puzzle 43

$$4 \times 5 + 3 = 23$$
$$+ \qquad + \qquad \times$$
$$7 + 2 + 9 = 18$$
$$\times \qquad \times \qquad +$$
$$6 + 8 + 1 = 15$$
$$= 46 \qquad = 21 \qquad = 28$$

Solution of Puzzle 44

$$3 \times 9 + 7 = 34$$
$$+ \qquad + \qquad \times$$
$$8 + 4 + 2 = 14$$
$$+ \qquad + \qquad \times$$
$$1 + 6 + 5 = 12$$
$$= 12 \qquad = 19 \qquad = 70$$

[121]

MATH IS FUN 5A

Solution of Puzzle 45

```
( 2 )+( 8 )x( 7 ) = 58
  x      x      +
( 1 )+( 5 )+( 3 ) = 9
  +      +      +
( 6 )x( 9 )+( 4 ) = 58
= 8    = 49    = 14
```

Solution of Puzzle 46

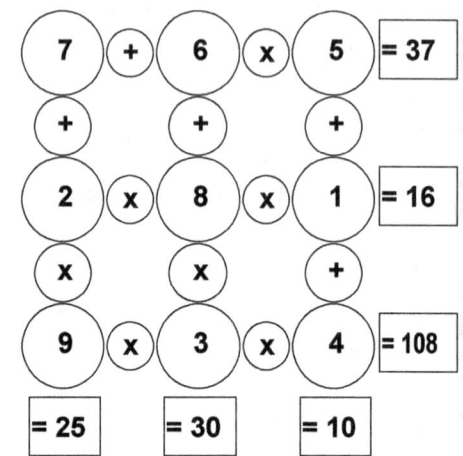

Solution of Puzzle 47

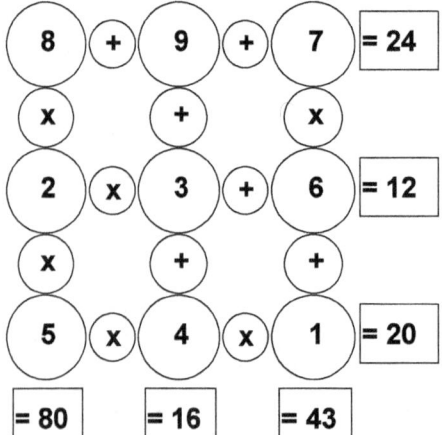

Solution of Puzzle 48

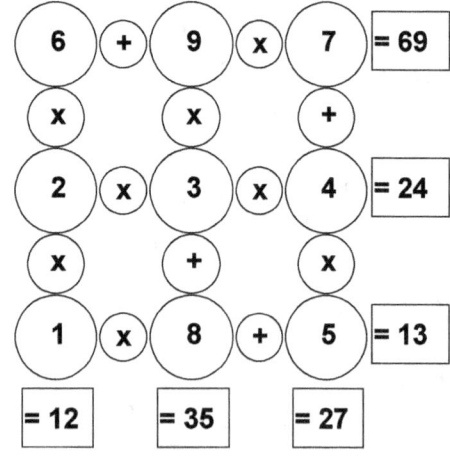

[122]

MATH IS FUN 5A

Solution of Puzzle 49

```
 4  +  5  x  3  = 19
 x     x     +
 9  x  1  +  8  = 17
 +     x     x
 2  x  6  x  7  = 84
= 38  = 30  = 59
```

Solution of Puzzle 50

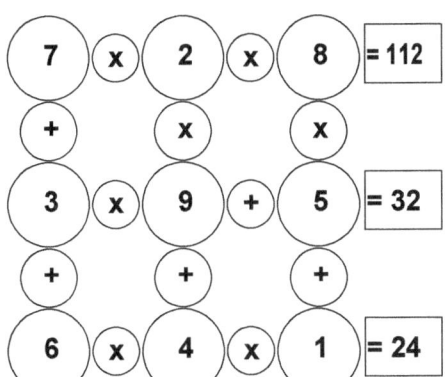

```
 7  x  2  x  8  = 112
 +     x     x
 3  x  9  +  5  = 32
 +     +     +
 6  x  4  x  1  = 24
= 16  = 22  = 41
```

Solution of Puzzle 51

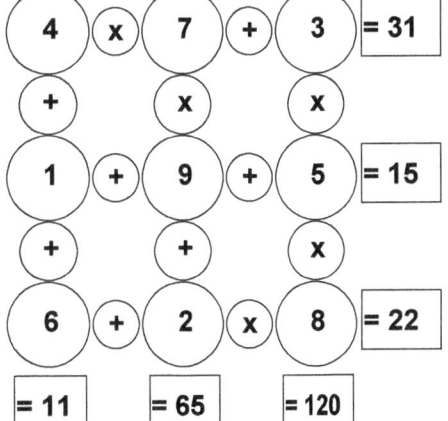

```
 4  x  7  +  3  = 31
 +     x     x
 1  +  9  +  5  = 15
 +     +     x
 6  +  2  x  8  = 22
= 11  = 65  = 120
```

Solution of Puzzle 52

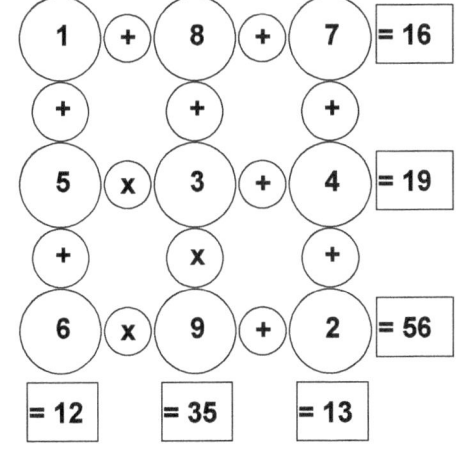

```
 1  +  8  +  7  = 16
 +     +     +
 5  x  3  +  4  = 19
 +     x     +
 6  x  9  +  2  = 56
= 12  = 35  = 13
```

[123]

MATH IS FUN 5A

Solution of Puzzle 53

9	+ 3	+ 4	= 16
x	+	+	
7	+ 1	x 2	= 9
+	+	x	
6	+ 8	x 5	= 46
= 69	= 12	= 14	

Solution of Puzzle 54

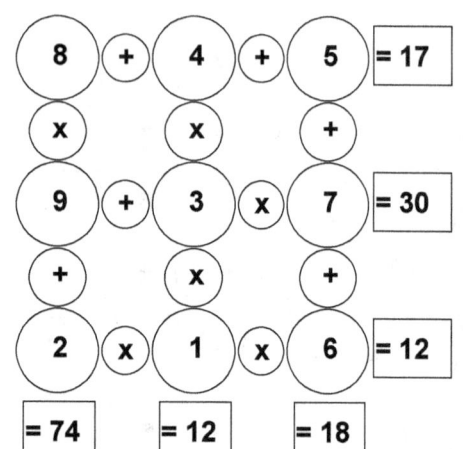

Solution of Puzzle 55

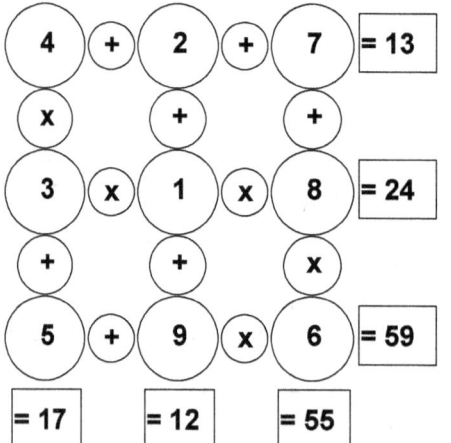

Solution of Puzzle 56

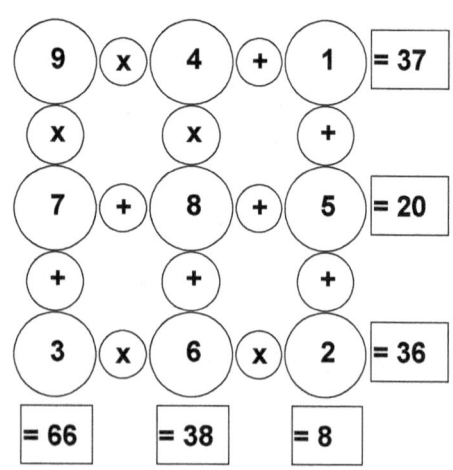

[124]

MATH IS FUN 5A

Solution of Puzzle 57

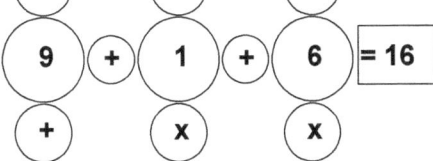
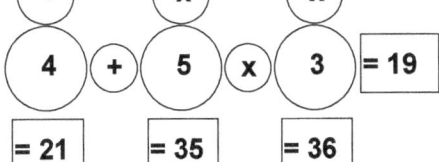

8 + 7 + 2	= 17
+ × ×	
9 + 1 + 6	= 16
+ × ×	
4 + 5 × 3	= 19
= 21	= 35	= 36

Solution of Puzzle 58

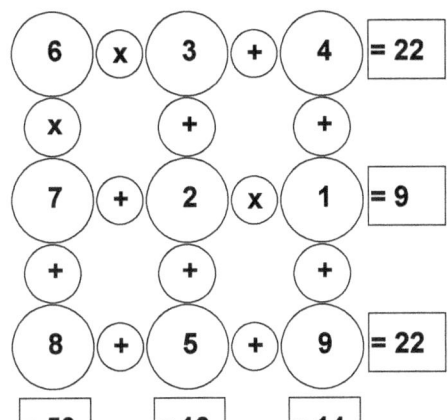

6 × 3 + 4	= 22
× + +	
7 + 2 × 1	= 9
+ + +	
8 + 5 + 9	= 22
= 50	= 10	= 14

Solution of Puzzle 59

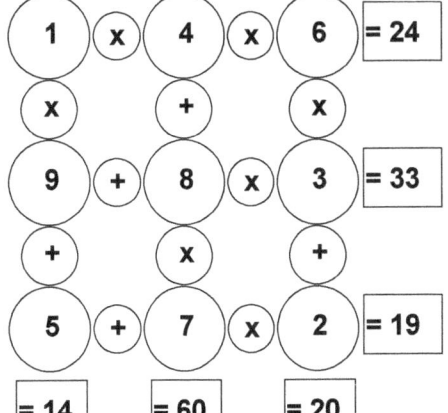

1 × 4 × 6	= 24
× + ×	
9 + 8 × 3	= 33
+ × +	
5 + 7 × 2	= 19
= 14	= 60	= 20

Solution of Puzzle 60

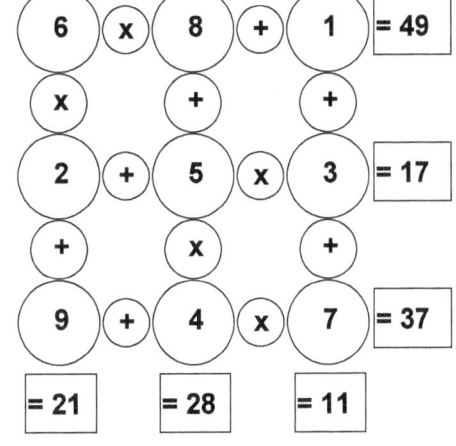

6 × 8 + 1	= 49
× + +	
2 + 5 × 3	= 17
+ × +	
9 + 4 × 7	= 37
= 21	= 28	= 11

[125]

MATH IS FUN 5A

Solution of Puzzle 61

$$4 \times 9 + 7 = 43$$
$$+ \quad \times \quad +$$
$$1 \times 6 \times 5 = 30$$
$$\times \quad + \quad +$$
$$3 \times 8 + 2 = 26$$
$$= 7 \qquad = 62 \qquad = 14$$

Solution of Puzzle 62

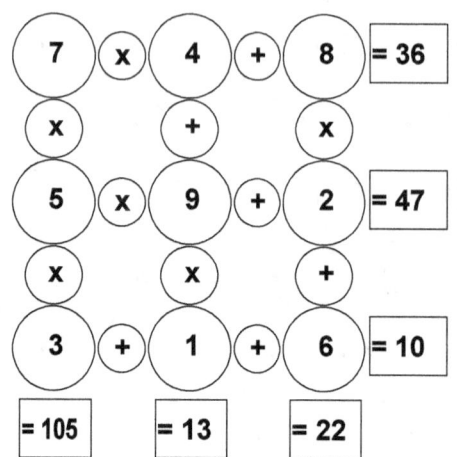

$$7 \times 4 + 8 = 36$$
$$\times \quad + \quad \times$$
$$5 \times 9 + 2 = 47$$
$$\times \quad \times \quad +$$
$$3 + 1 + 6 = 10$$
$$= 105 \qquad = 13 \qquad = 22$$

Solution of Puzzle 63

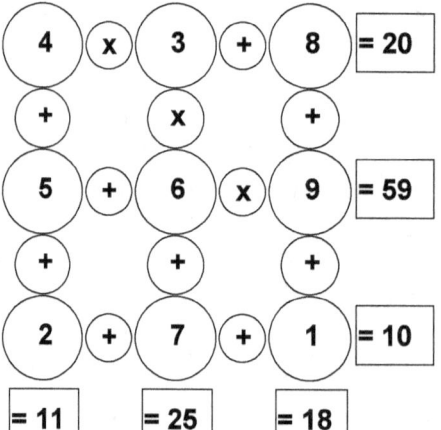

$$4 \times 3 + 8 = 20$$
$$+ \quad \times \quad +$$
$$5 + 6 \times 9 = 59$$
$$+ \quad + \quad +$$
$$2 + 7 + 1 = 10$$
$$= 11 \qquad = 25 \qquad = 18$$

Solution of Puzzle 64

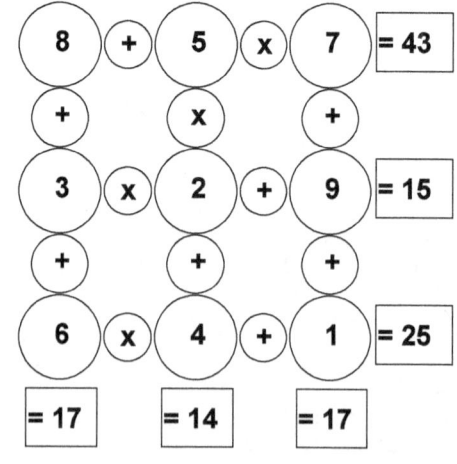

$$8 + 5 \times 7 = 43$$
$$+ \quad \times \quad +$$
$$3 \times 2 + 9 = 15$$
$$+ \quad + \quad +$$
$$6 \times 4 + 1 = 25$$
$$= 17 \qquad = 14 \qquad = 17$$

[126]

MATH IS FUN 5A

Solution of Puzzle 65

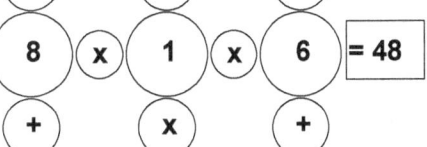

4	x	5	x	3	= 60

4 × 5 × 3 = 60

+ , + , x

8 × 1 × 6 = 48

+ , x , +

2 + 7 + 9 = 18

= 14 , = 12 , = 27

Solution of Puzzle 66

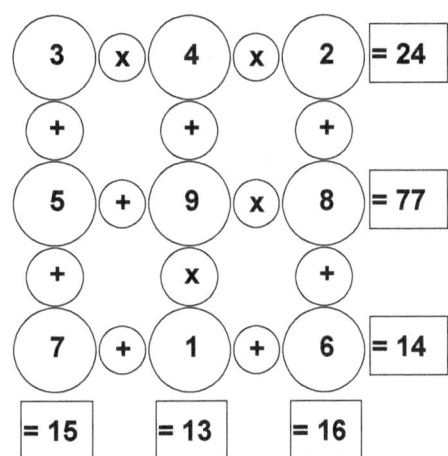

3 × 4 × 2 = 24

+ , + , +

5 + 9 × 8 = 77

+ , x , +

7 + 1 + 6 = 14

= 15 , = 13 , = 16

Solution of Puzzle 67

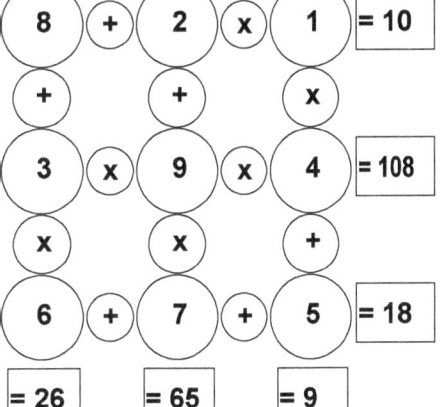

8 + 2 × 1 = 10

+ , + , x

3 × 9 × 4 = 108

x , x , +

6 + 7 + 5 = 18

= 26 , = 65 , = 9

Solution of Puzzle 68

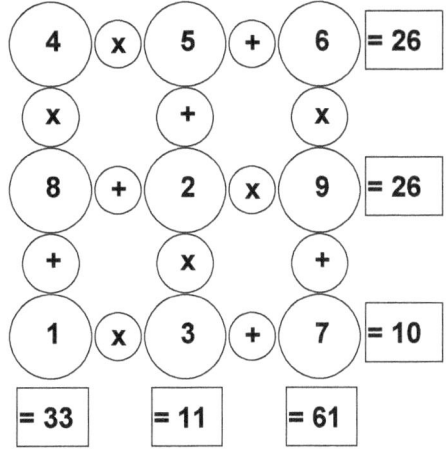

4 × 5 + 6 = 26

x , + , x

8 + 2 × 9 = 26

+ , x , +

1 × 3 + 7 = 10

= 33 , = 11 , = 61

[127]

MATH IS FUN 5A

Solution of Puzzle 69

```
( 3 ) + ( 1 ) x ( 4 ) = 7
  +       +       +
( 2 ) x ( 9 ) + ( 6 ) = 24
  +       +       +
( 5 ) + ( 8 ) x ( 7 ) = 61
= 10    = 18    = 17
```

Solution of Puzzle 70

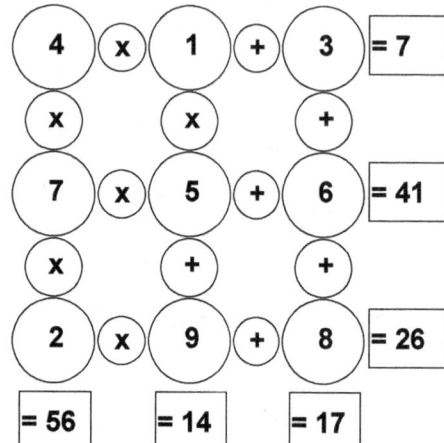

```
( 4 ) x ( 1 ) + ( 3 ) = 7
  x       x       +
( 7 ) x ( 5 ) + ( 6 ) = 41
  x       +       +
( 2 ) x ( 9 ) + ( 8 ) = 26
= 56    = 14    = 17
```

Solution of Puzzle 71

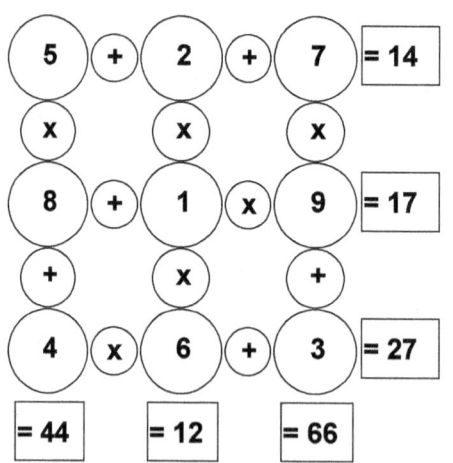

```
( 5 ) + ( 2 ) + ( 7 ) = 14
  x       x       x
( 8 ) + ( 1 ) x ( 9 ) = 17
  +       x       +
( 4 ) x ( 6 ) + ( 3 ) = 27
= 44    = 12    = 66
```

Solution of Puzzle 72

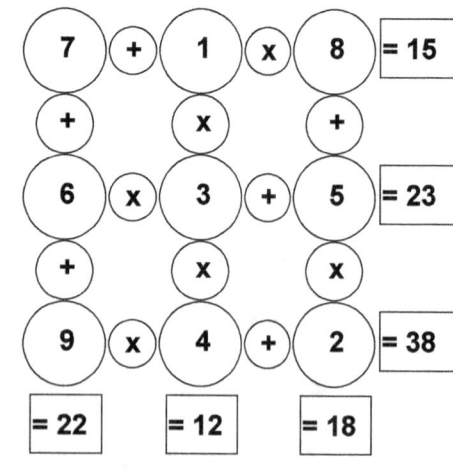

```
( 7 ) + ( 1 ) x ( 8 ) = 15
  +       x       +
( 6 ) x ( 3 ) + ( 5 ) = 23
  +       x       x
( 9 ) x ( 4 ) + ( 2 ) = 38
= 22    = 12    = 18
```

[128]

MATH IS FUN 5A

Solution of Puzzle 73

```
( 1 ) x ( 9 ) x ( 2 ) = 18
  +       x       x
( 5 ) x ( 8 ) + ( 4 ) = 44
  +       +       +
( 6 ) x ( 3 ) + ( 7 ) = 25
= 12    = 75    = 15
```

Solution of Puzzle 74

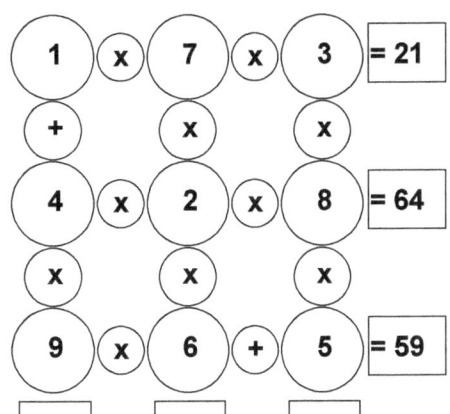

```
( 1 ) x ( 7 ) x ( 3 ) = 21
  +       x       x
( 4 ) x ( 2 ) x ( 8 ) = 64
  x       x       x
( 9 ) x ( 6 ) + ( 5 ) = 59
= 37    = 84    = 120
```

Solution of Puzzle 75

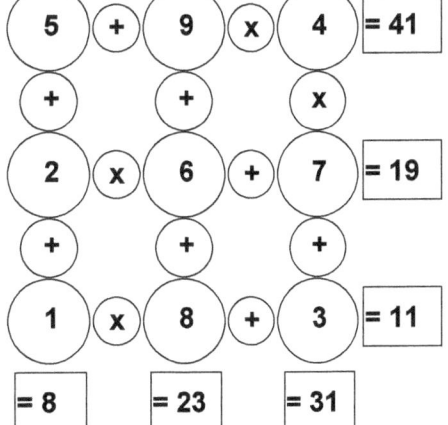

```
( 5 ) + ( 9 ) x ( 4 ) = 41
  +       +       x
( 2 ) x ( 6 ) + ( 7 ) = 19
  +       +       +
( 1 ) x ( 8 ) + ( 3 ) = 11
= 8     = 23    = 31
```

Solution of Puzzle 76

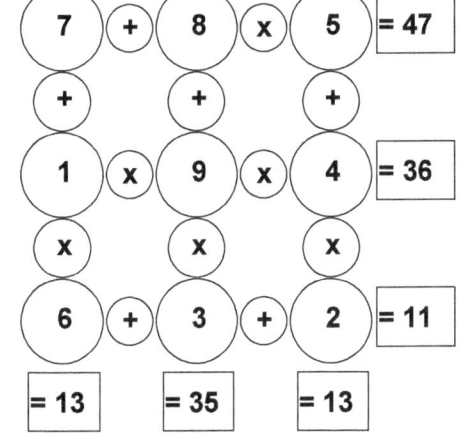

```
( 7 ) + ( 8 ) x ( 5 ) = 47
  +       +       +
( 1 ) x ( 9 ) x ( 4 ) = 36
  x       x       x
( 6 ) + ( 3 ) + ( 2 ) = 11
= 13    = 35    = 13
```

[129]

MATH IS FUN 5A

Solution of Puzzle 77

```
(7) x (1) + (2) = 9
 x       x       +
(9) + (8) x (5) = 49
 +       x       x
(3) + (6) + (4) = 13
= 66    = 48    = 22
```

Solution of Puzzle 78

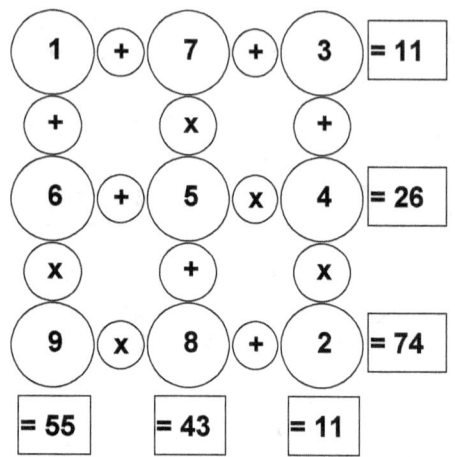

```
(1) + (7) + (3) = 11
 +       x       +
(6) + (5) x (4) = 26
 x       +       x
(9) x (8) + (2) = 74
= 55    = 43    = 11
```

Solution of Puzzle 79

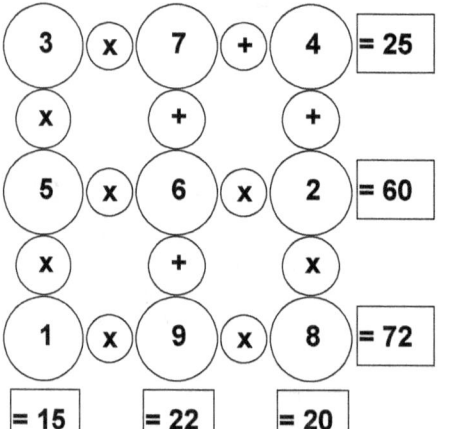

```
(3) x (7) + (4) = 25
 x       +       +
(5) x (6) x (2) = 60
 x       +       x
(1) x (9) x (8) = 72
= 15    = 22    = 20
```

Solution of Puzzle 80

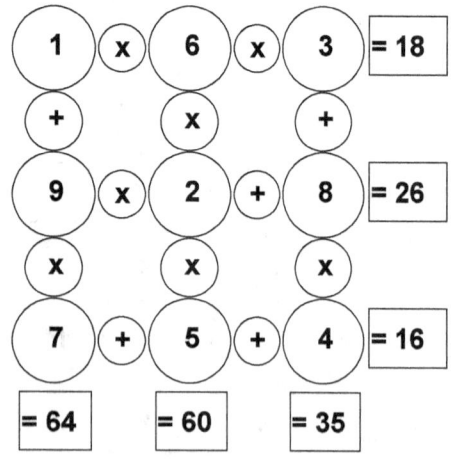

```
(1) x (6) x (3) = 18
 +       x       +
(9) x (2) + (8) = 26
 x       x       x
(7) + (5) + (4) = 16
= 64    = 60    = 35
```

MATH IS FUN 5A

Solution of Puzzle 81

```
8  +  3  x  5  = 23
+     +     +
6  x  4  +  7  = 31
+     +     x
2  x  1  x  9  = 18
= 16  = 8   = 68
```

Solution of Puzzle 82

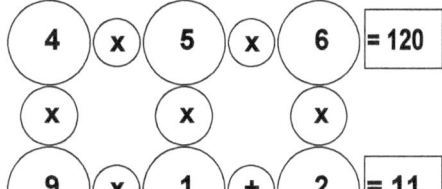
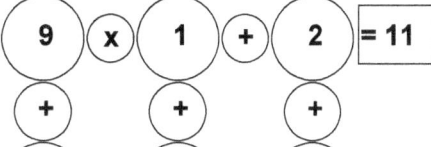

```
4  x  5  x  6  = 120
x     x     x
9  x  1  +  2  = 11
+     +     +
3  x  8  +  7  = 31
= 39  = 13  = 19
```

Solution of Puzzle 83

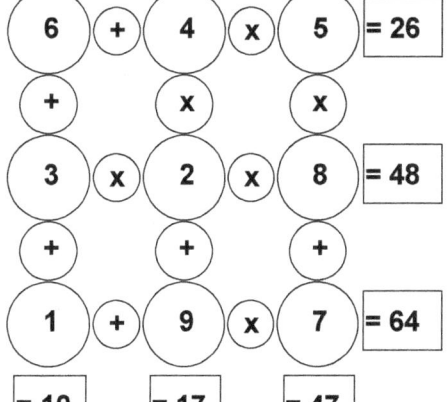

```
6  +  4  x  5  = 26
+     x     x
3  x  2  x  8  = 48
+     +     +
1  +  9  x  7  = 64
= 10  = 17  = 47
```

Solution of Puzzle 84

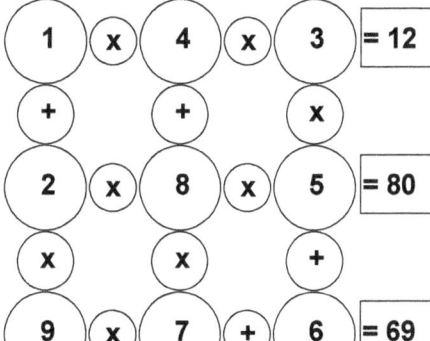

```
1  x  4  x  3  = 12
+     +     x
2  x  8  x  5  = 80
x     x     +
9  x  7  +  6  = 69
= 19  = 60  = 21
```

MATH IS FUN 5A

Solution of Puzzle 85

2	x	8	+	4		= 20
+		x		+		
3	+	6	+	7		= 16
+		+		x		
9	x	1	+	5		= 14
= 14		= 49		= 39		

Solution of Puzzle 86

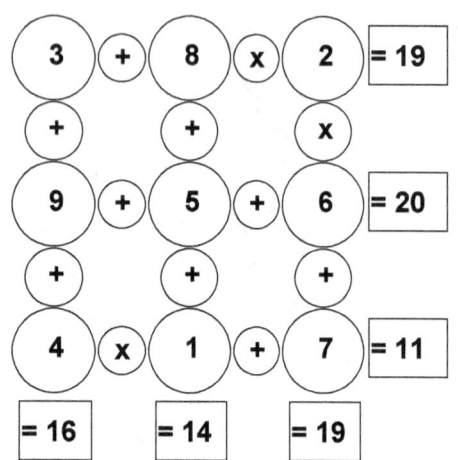

3	+	8	x	2		= 19
+		+		x		
9	+	5	+	6		= 20
+		+		+		
4	x	1	+	7		= 11
= 16		= 14		= 19		

Solution of Puzzle 87

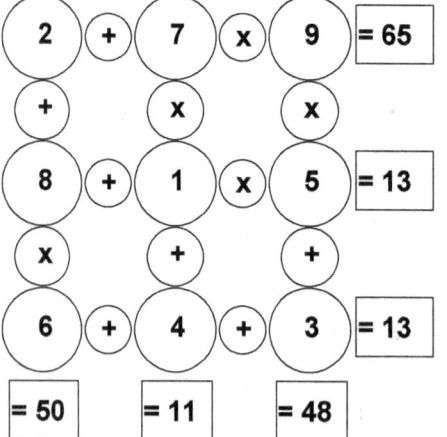

2	+	7	x	9		= 65
+		x		x		
8	+	1	x	5		= 13
x		+		+		
6	+	4	+	3		= 13
= 50		= 11		= 48		

Solution of Puzzle 88

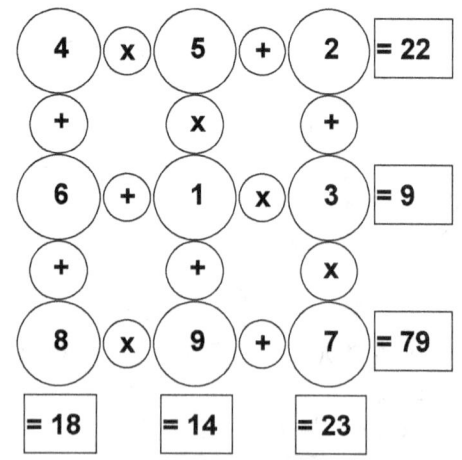

4	x	5	+	2		= 22
+		x		+		
6	+	1	x	3		= 9
+		+		x		
8	x	9	+	7		= 79
= 18		= 14		= 23		

MATH IS FUN 5A

Solution of Puzzle 89

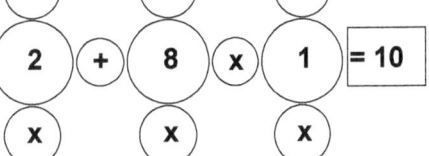
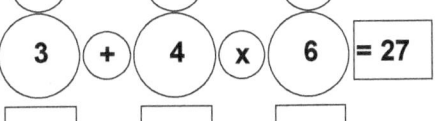

```
 7  x  5  + 9  = 44
 x     +     +
 2  +  8  x 1  = 10
 x     x     x
 3  +  4  x 6  = 27
=42   =37   =15
```

Solution of Puzzle 90

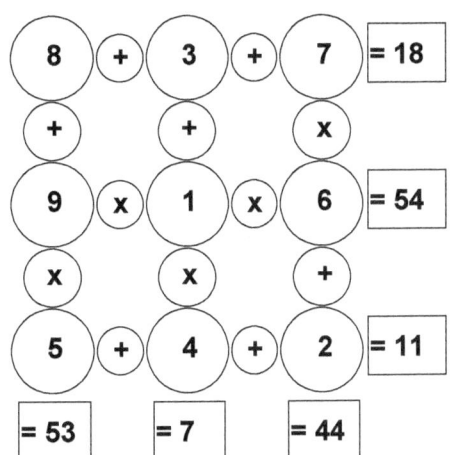

```
 8  +  3  + 7  = 18
 +     +     x
 9  x  1  x 6  = 54
 x     x     +
 5  +  4  + 2  = 11
=53   =7    =44
```

Solution of Puzzle 91

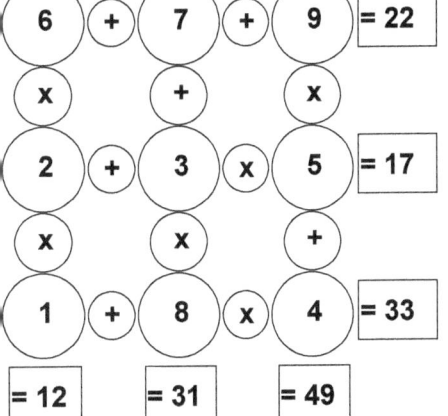

```
 6  +  7  + 9  = 22
 x     +     x
 2  +  3  x 5  = 17
 x     x     +
 1  +  8  x 4  = 33
=12   =31   =49
```

Solution of Puzzle 92

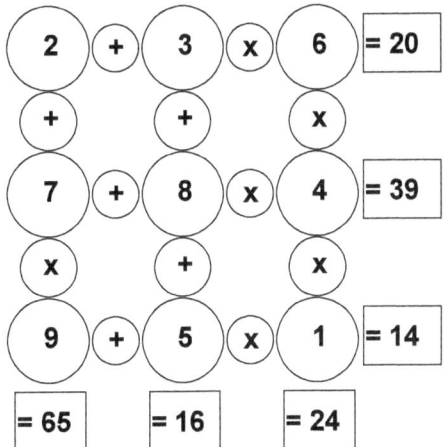

```
 2  +  3  x 6  = 20
 +     +     x
 7  +  8  x 4  = 39
 x     +     x
 9  +  5  x 1  = 14
=65   =16   =24
```

[133]

MATH IS FUN 5A

Solution of Puzzle 93

```
( 2 ) + ( 5 ) x ( 6 ) = 32
  +       +       +
( 3 ) + ( 7 ) x ( 8 ) = 59
  x       +       x
( 9 ) x ( 1 ) x ( 4 ) = 36
= 29    = 13    = 38
```

Solution of Puzzle 94

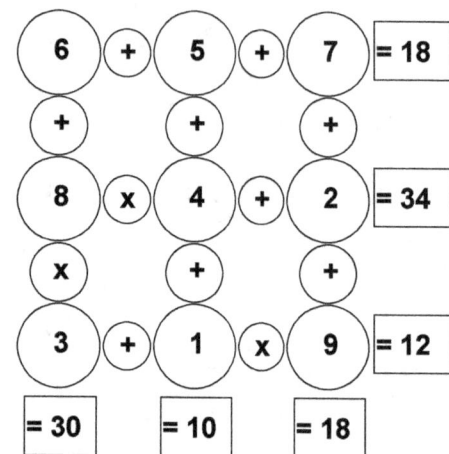

```
( 6 ) + ( 5 ) + ( 7 ) = 18
  +       +       +
( 8 ) x ( 4 ) + ( 2 ) = 34
  x       +       +
( 3 ) + ( 1 ) x ( 9 ) = 12
= 30    = 10    = 18
```

Solution of Puzzle 95

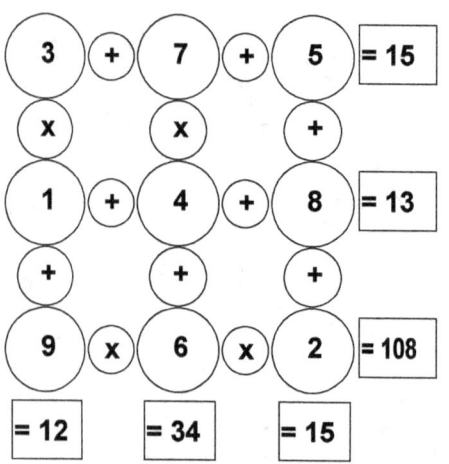

```
( 3 ) + ( 7 ) + ( 5 ) = 15
  x       x       +
( 1 ) + ( 4 ) + ( 8 ) = 13
  +       +       +
( 9 ) x ( 6 ) x ( 2 ) = 108
= 12    = 34    = 15
```

Solution of Puzzle 96

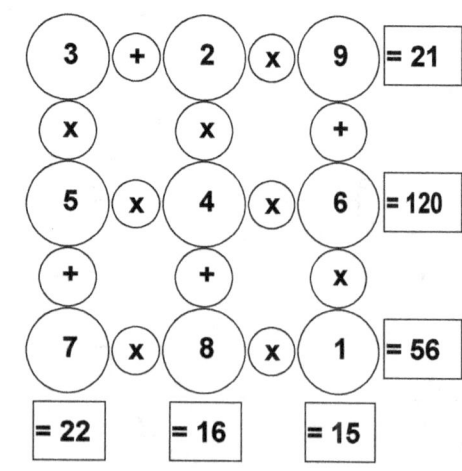

```
( 3 ) + ( 2 ) x ( 9 ) = 21
  x       x       +
( 5 ) x ( 4 ) x ( 6 ) = 120
  +       +       x
( 7 ) x ( 8 ) x ( 1 ) = 56
= 22    = 16    = 15
```

[134]

MATH IS FUN 5A

Solution of Puzzle 97

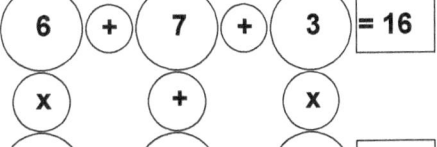

4 + 2 + 1	= 7
+ + +	
6 + 7 + 3	= 16
× + ×	
8 + 5 × 9	= 53
= 52	= 14	= 28

Solution of Puzzle 98

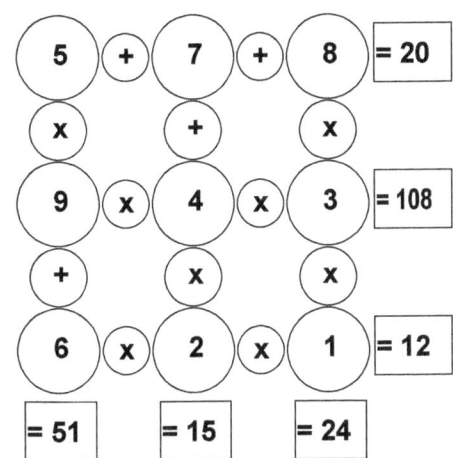

5 + 7 + 8	= 20
× + ×	
9 × 4 × 3	= 108
+ × ×	
6 × 2 × 1	= 12
= 51	= 15	= 24

Solution of Puzzle 99

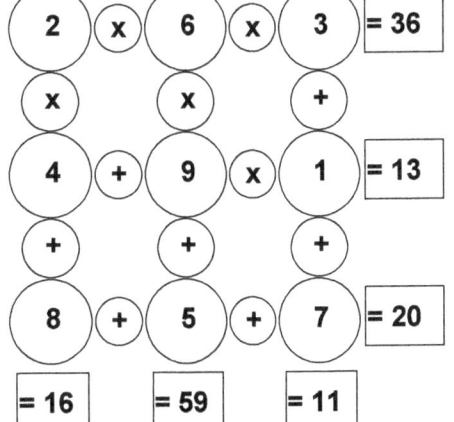

2 × 6 × 3	= 36
× × +	
4 + 9 × 1	= 13
+ + +	
8 + 5 + 7	= 20
= 16	= 59	= 11

Solution of Puzzle 100

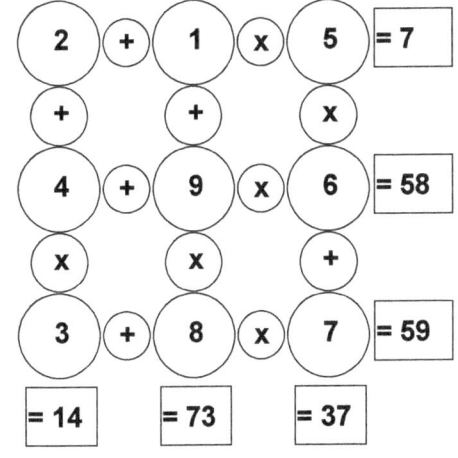

2 + 1 × 5	= 7
+ + ×	
4 + 9 × 6	= 58
× × +	
3 + 8 × 7	= 59
= 14	= 73	= 37

www.ingramcontent.com/pod-product-compliance
Lightning Source LLC
Chambersburg PA
CBHW071315220526
45468CB00001B/383